# Asset Management Decision-Making For Infrastructure Systems

Alireza Mohammadi • Luis Amador Jimenez

# Asset Management Decision-Making For Infrastructure Systems

Springer

Alireza Mohammadi
Postdoctoral Fellow
University of Toronto
Toronto, ON, Canada

Luis Amador Jimenez
Concordia University
Montreal, QC, Canada

ISBN 978-3-030-97613-2 ISBN 978-3-030-97614-9 (eBook)
https://doi.org/10.1007/978-3-030-97614-9



This Springer imprint is published by the registered company Springer Nature Switzerland AG
The registered company address is: Gewerbestrasse 11, 6330 Cham, Switzerland

# Preface

Back in 2007, I (Dr. Luis Amador) was confronted with the daunting task of developing a performance deterioration model and decision support system for pavements of the Costa Rica national road network from only two condition surveys, at the time I was studying for my Ph.D. and working on the implementation of the Transportation Asset Management System for New Brunswick Department of Transportation. For nearly 3 years, I developed performance deterioration models, studied mathematical algorithms for optimization of decisions, and met regularly to help guide the write-up of an implementation model and change management practices for infrastructure management organizations. I quickly identified multiple limitations such as insufficient time-series data to develop deterioration models or to estimate the effectiveness of maintenance and rehabilitation in order to count with powerful servers and to solve an optimization model. In general, the lack of guidelines to develop asset management decision support models.

Soon after finalizing my studies, I was engaged by the Town of Kindersley that had coincidentally embarked on the development of municipal infrastructure management systems. This brought a whole new dimension and challenges. First, there was no budget to pay for a multifunctional vehicle to come do a condition survey. Second, there was an imperative need to coordinate urgent interventions on very aged pipes (water mains, sanitary sewers, and storm pipes) with road rehabilitation and reconstruction. For once, there were old paper records of historical interventions. I left Kindersley for a job at Concordia University with the promise in my head to find solutions. Time passed and more challenges arrived in the form of having road management systems capable of handling road safety goals to reduce the severity and frequency of crashes by providing remedial works and safety hardware. By 2013, I was working for the government of Ecuador in South America to develop a road management system for the province of El Oro. For 2 years, we organized their road management system, met with stakeholders, and led a campaign to collect roughness data with mobile devices due to the lack of resources to bring standard survey condition equipment. The province had only one university and seven hospitals spread through their territory, and travel from some regions to a hospital or work in the city would take hours. We were confronted with the need to consider travel time as part of the goals for scheduling upgrade of gravel roads or realign old trails. One of the pending issues in my head was the suspicion that historical traffic growth is inadequate to forecast future truck traffic growth for infrastructure asset management.

In 2016, a brilliant Ph.D. student, Mr. Alierza Mohammadi, with a strong industry background in infrastructure engineering and construction joined my team. We started working on the development of innovative and practical infrastructure management solutions for roads, railway transits, stormwater systems, and municipal assets. Alireza focused on urban railway systems, where multiple challenges appeared. The stations are systems with multiple subsystems, and the users value the convenience and comfort of the service in addition to cost, travel time, and safety. Transit systems are capable of facilitating access to employment, education, and health. Therefore, decisions on the reactivation of old rails or the expansion of transit systems must consider such benefits.

Alireza graduated and moved quickly to join the University of Toronto as a postdoctoral fellow, to deal with practical asset management projects from governments and municipalities. He also led to exploring new aspects of infrastructure management systems to develop a user-oriented platform comparing classical asset-oriented approaches. Later, he joined Metrolinx, a government agency building and operating transit systems across the Greater Toronto Area, to implement advanced asset management solutions.

We see how standards and guidelines for asset management continue to be good abstractions, providing high-level understanding; however, they are incomprehensible to those putting their hands on the actual development of decision support systems and associated models. All standards and guidelines highly recommend planning for asset management based on optimized decision-making; however, the key questions for professionals are which steps should be taken and what are the requirements to achieve optimum plans and budget allocations. This book fills the technical gaps between these guidelines and practical needs by providing and presenting step-by-step instructions derived from state-of-the-art solutions and already-implemented industry best practices.

By 2019, we realized the necessity to compile these concrete learnings from case studies and consulting works as well as the need to document all these practical advances in one book to provide concrete guidance to students and professionals, and so we proposed to write a book. Today, dear reader, we hand over the product of over 10 years of research and consulting works to you with the hopes it will facilitate your work, and with the conviction that the work of infrastructure professionals contributes to the livelihood of our cities and the well-being of its inhabitants.

Toronto, ON, Canada — Alireza Mohammadi
Montreal, QC, Canada — Luis Amador Jimenez

# Contents

# Chapter 1
# Introduction to Asset Management for Infrastructure

## 1.1 Infrastructure

Oxford dictionary (2020) defines infrastructure as "the basic physical and organizational structures and facilities (e.g., buildings, roads, power supplies) needed for the operation of a society or enterprise.

A more technical definition has been provided by the American Society of Civil Engineering (ASCE): Civil infrastructure systems enable thriving societies and healthy ecosystems. Civil infrastructure systems support transportation; energy production and distribution; water resources management; waste management; civic facilities in urban and rural communities; communications; sustainable resources development; and environmental protection. These physical, social, ecological, economic, and technological systems are complex and interrelated" (ASCE, 2021).

Infrastructure UK encompasses economic and system perspectives to explain infrastructure roles: "the networks and systems in energy, transport, digital communication, flood protection, water, and waste management. These are all critical to support economic growth through the expansion of private sector businesses across all regions and industries, to enable competitiveness and to improve the quality of life of everyone in the UK" (Treasury, 2010).

Most references agree on similar major sectors as Critical Infrastructure (CI) including transport; energy; water and wastewater; and communication systems, which typically are categorized as "hard" infrastructure referring to physical assets. ASCE (2017) expanded this to 16 categories of Transit, Aviation, Bridges, Roads, Ports, Rail, Energy, Inland Waterways, Dams, Drinking Water, Wastewater, Solid Waste, Hazardous Waste, Levees, Schools, and Parks and Recreation. Green infrastructures such as food, agriculture, and chemicals as well as soft infrastructures such as health, education, and legal systems are also included by other sources (Dawson, 2013).



A. Mohammadi, L. Amador Jimenez, *Asset Management Decision-Making For Infrastructure Systems*, https://doi.org/10.1007/978-3-030-97614-9_1

Among all different views and definitions for hard infrastructure, there are key common, agreed aspects that are central to the role of infrastructure in modern societies and are the pillars of the treatment in this book.

### 1.1.1 Physical Bases

Infrastructure is commonly referred to as physical components, structures, and assets, which can be built, installed, repaired, and replaced. Infrastructure assets are touchable; however, the commodities that flow through them such as oil and gas or electric power are not part of the system although may physically or financially impact the system (Fulmer, 2009).

### 1.1.2 Criticality

People's human life and well-being are highly dependent on the major infrastructure, commonly named CI. In many societies, it would not be possible to imagine a single day without using electric power, transportation, and communication systems. The government of Canada (2020) says that "CI refers to processes, systems, facilities, technologies, networks, assets and services essential to the health, safety, security or economic well-being of Canadians and the effective functioning of government. Disruptions of CI could result in catastrophic loss of life, adverse economic effects, and significant harm to public confidence." The criticality of major services such as food supply, electricity grids, transportation, communications, and public safety incorporates managing infrastructure to cybersecurity and resiliency.

### 1.1.3 Economic and Human Development

Investing in infrastructure is always very expensive, however globally is recognized as a key source of running the business and facilitating economic growth. One main requirement for sustainable economic and social development is providing adequate infrastructure services (UN, 1994).

Identified Sustainable Development Goals (SDGs) by the United Nations (UN, 2016) including access to safe and reliable transport; clean water and sanitation; affordable and clean energy; quality education; good health and well-being; and empowerment of women, persons with disabilities, and other vulnerable groups are directly influenced by the quality of provided services by the infrastructure.

### 1.1.4 *Public Facilities*

Major infrastructure is often monopolistic in terms of provided facilities (Fulmer, 2009), and the governments and municipalities are responsible to invest, maintaining, and upkeeping these services. Thus, the term "Public Infrastructure" most likely covers all CI as is defined by the Corporate Finance Institute (CFI): "Public infrastructure refers to infrastructure facilities, systems, and structures that are owned and operated by the "public," (i.e., the government). It includes all infrastructural facilities that are open to the general public for use. Infrastructure includes all essential systems and facilities that facilitate the smooth flow of an economy's day-to-day activities and enhance the people's standard of living. It includes basic facilities such as roads, water supply, electricity, and telecommunications (CFI, 2020)".

## 1.2 Infrastructure Network

Our societies are built around and served by infrastructure networks. Major systems such as transportation, energy, and communication have mostly been managed individually; however, infrastructure now functions as a system of systems with complex interdependencies (Grafius et al., 2020). Infrastructure systems or networks are interrelated components functioning together in a dependent system in order to deliver services to end-users. Several interdependencies can be identified including supply and demand (e.g., water system relies on electricity); physical dependencies (e.g., installing the antenna on a building or buried utilities under a road); geographic dependency (e.g., close spatial proximity); economic dependencies (e.g., investment cycles for each system); and finally, technological dependency (e.g., managing highway traffic by sensors) (Fulmer, 2009) (Fig. 1.1).

Widespread use of information across infrastructure systems, future reliance on communication technology (e.g., vehicle to infrastructure and V2i technology), and integration of technology have improved the efficiency of infrastructure while dropping resilience and increasing the chance of systemic failure (i.e., technology vulnerability) (Treasury, 2010).

Provided services by infrastructure networks are firstly dependent on the quality, reliability, and availability of each system; however, from a network perspective, managing infrastructure is beyond each system considering interdependencies and addressing the network resiliency, which can be also influenced by losing functionality due to abnormal failure (e.g., natural thread as well as terrorist attacks). CI protection is usually discussed separately addressing several system challenges such as cybersecurity, which is out of the scope of this book. More discussions can be found in several sources such as the *International Journal of Critical Infrastructure Protection (IJCIP)* (Elsevier, 2020).

Fig. 1.1 Infrastructure facilities interdependencies

## 1.3 Infrastructure Asset Management (IAM)

Managing infrastructure assets especially public services is a complex and multi-dimensional task touching commercial, social, economic, and political aspects. Cambridge dictionary defines an asset as "something having value, such as a possession or property, that is owned by a person, business, or organization" (Cambridge, 2020). The ISO (2014a) defines an asset as "an item, thing, or entity that has potential or actual value to an organization". Therefore, different sorts of assets including physical, financial, human, information, and intangible assets can be identified within any organization, while physical assets in infrastructure refer to tangible elements of buildings, roads, pipelines, plants, and communication equipment (Hastings, 2015). Assets in the context of this book are those physical components and systems.

All organizations that manage our infrastructure such as governments, municipalities, and all private parties own, maintain, and deal with assets that drive the quality of provided services. To assure quality and availability of expected services, assets should be designed, installed, built, operated, repaired, replaced, and upgraded, and overall managed properly in order to return the highest value to the owners as well as end-users. This is not new for industry; however, the growing responsibility of organizations, increasing in number and complexity of assets, operating under a restricted budget, and rising demands lead to the need to use a systematic approach to maintain assets and systems called asset management (AM).

The Federation of Canadian Municipalities (FCM) and National Research Council (NRC) (2005) defined AM as "the combination of management, financial, economic, engineering, and other practices applied to physical assets with the

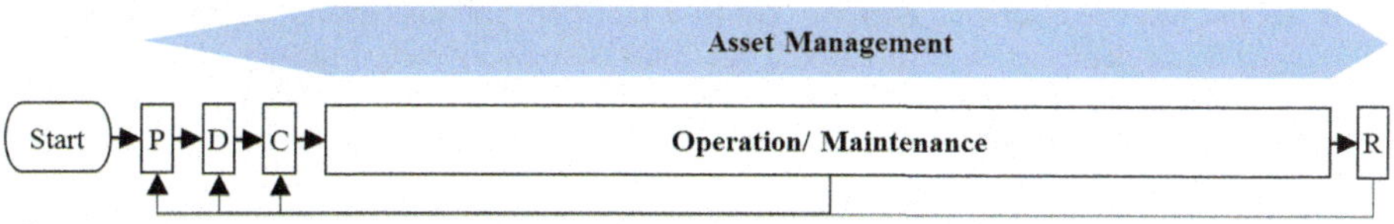


**Fig. 1.2** Asset life cycle and AM

objective of providing the required level of service in the most cost-effective manner." United States Army Corps of Engineers (USACE) the Institute for Water Resources (IWR) (2013) explained AM as a disciplined corporate approach requiring collaboration at the organization level.

Figure 1.2 shows the asset life cycle: starting from Planning (P) and Design (D), moving to Construction and Commissioning (C), then Operation (O), which assets spend the longest time up to Replacement and Retirement (R), where the asset has completed its useful life and will be decommissioned. AM mainly focuses on the operation and maintenance phase while contributing to other phases and providing feedback to the entire asset life cycle. For instance, how the decommissioned asset should be redesigned, reinstalled, or handed over to improve the service in the operation phase. AM helps an organization to maintain assets proactively and efficiently by finding the optimum time and strategy for the maintenance and replacement of assets.

All organizations may establish their own understanding of AM; however, to achieve the highest benefit, it would be highly recommended to follow best practices as a baseline and then begin from that point to address local concerns through adjustments.

Davis (2012) introduced AM for beginners in a simple way of what it is and what it is not. AM is:

- "A mindset which sees physical assets not as inanimate and unchanging lumps of metal/plastic/concrete, but as objects and systems which respond to their environment, change and normally deteriorate with use, and progressively grow old then fail, stop working, and eventually die!
- **Is** a recognition that assets have a life cycle
- **Is** as important for those working in finance as it is for engineers
- **Is** an approach that looks to get the best out of the assets for the benefit of the organization and/or its stakeholders
- **Is** about understanding and managing the risk associated with owning assets

And

- **Is not** just about maintenance. Maintenance is part of the stewardship of assets, but so is design, procurement, installation, commissioning, operation, etc.
- **Is not** a substitute for quality management. AM, like other management processes, should be subject to scrutiny through a quality process to ensure rigour.
- **Is not** a project management system
- **Is not** just for engineers. Everyone working in a company that owns or operates assets should be interested. This includes those working in procurement, finance, personnel, service, planning, design, operations, administration, leadership, marketing, and sales

- **Is not** just an accounting exercise. Whilst it may help you understand the deterioration and hence depreciation of an asset, it is of interest to every part of the organization
- **Is not** a purely academic discipline. Whilst it is a worthy subject for academic review and advancement, it is primarily a pragmatic, hands-on subject".

Historically, AM has been first applied a long time ago in manufacturing systems with the main objective of avoiding any shutdown in the production line. Then, the idea was expanded to infrastructure, firstly applying to industrial-based systems such as oil and gas and later to civil infrastructure. Over time, IAM became more critical where infrastructure such as transit systems and buildings became bigger, older, more complex, and more financially dependent.

Let us terminate this section with another comprehensive asset management definition provided by the American Association of State Highway and Transportation Officials (AASHTO) Standing Committee on Highways, Planning Subcommittee on Asset Management (FHWA, 2021): "a strategic and systematic process of operating, maintaining, upgrading, and expanding physical assets effectively throughout their life cycle. It focuses on business and engineering practices for resource allocation and utilization, with the objective of better decision making based on quality information and well-defined objectives."

### *1.3.1 Why Asset Management (AM) for Infrastructure?*

Implementing AM for infrastructure in public or private organizations is time-consuming and requires financial resources and sufficient personnel. It often involves an action plan to build up organizational capabilities, change management to accomplish buy-in from the employees, and the development of enhanced information technology systems. Therefore, it must be worth an investment in the eyes of asset owners and policymakers.

Answers to the two key questions of why we need IAM and what would be the benefits of this investment should be well prepared and articulated, particularly for senior managers.

Below are some of the reasons why an organization may be willing to develop and implement an AM platform:

- Community well-being is paramount and properly linked to infrastructure availability and services reliability.
- There is a desire to optimize the management of local infrastructure systems, which are big and more complex in nature.
- There is a need to handle existent major infrastructure, which is suffering from aging while is under operation.
- The annual budget is limited, and this is perceived as an obstacle to upkeep assets in good operational condition.
- Increased demand because of the growing population and urbanization is observed and/or expected.

- Higher standards for lean operation, safety, and health are expected by end-users.
- Environmental protection concerns and sustainability issues are a top priority for the stakeholders.

This situation brings a need to maintain and upgrade existing infrastructure assets addressing challenges, which would not be achievable unless putting a systematic approach for managing assets. AM is intended to build this systematic practice.

### *1.3.2 Benefits of IAM*

IAM can work as a process to consistently and actively maintain infrastructure and to achieve a state of good repair (SGR). The following benefits can be gained in different degrees:

- Providing better and consistent levels of service for the customers.
- Improving safety, security, customer satisfaction, sustainability, and resiliency.
- Allowing for better data-driven, smart, and optimal decisions when allocating maintenance, rehabilitation, upgrade, and expansion of IAM systems.
- A reduction in cost during the life cycle of infrastructure and an increase in their life span.
- Facilitating more effective financial planning with more cost-effective use of resources.
- Avoiding service disruptions and minimizing potential risks.

### *1.3.3 State of Good Repair*

Federal Transportation Administration (FTA) (FTA, 2010) defines state of good repair (SGR) as "a state in which a transit agency preserves its physical assets in compliance with a policy that minimizes asset life-cycle costs while preventing adverse consequential impacts to its service." SGR actually is a bridge that links IAM practices to organization strategic objectives. Predefined SGR measures such as performance and condition indicators as well as the age of assets define objectives and targets to build AM platform and to ultimately achieve organizational key performance indicators (KPIs).

This book further develops an understanding of IAM strategic objectives through the challenges presented in Sect. 1.4; performance assessment (Chap. 2) and performance forecast (Chaps. 3 and 4); as well as the decision support platforms (Chap. 5) needed to accomplish the minimization of asset life cycle cost and the maximization of organization KPIs.

### *1.3.4 A Multi-discipline Process*

IAM is not either engineering or finance. It is a multi-disciplinary process supported by various knowledge areas (Fig. 1.3).

#### 1.3.4.1 Management

AM is all about management. The process should be built on management pillars including strategy, goals, policies, cost, time, quality, and definitely people.

#### 1.3.4.2 Engineering

Infrastructures are engineering systems where operation and maintenance are tied to engineering aspects of the asset's life cycle. Choosing the right treatment action at the right time is directly linked to engineering practices.

**Fig. 1.3** Multi-disciplinary contributions in IAM

#### 1.3.4.3 Economics

Infrastructure runs economics in any society while is expensive to be maintained. Meanwhile, governments are economically struggling to invest in infrastructure as the public side of the service limits making high revenue. Therefore, minimizing lifecycle costs and economically optimizing restricted budgets is critical.

#### 1.3.4.4 Sociology

Dependency on provided services by infrastructure systems and their influence on the quality of life are undeniable for societies. Besides that, IAM means dealing with under-operation systems. This requires considering all side effects on day to day life of end-users.

## 1.4 IAM Challenges

Although IAM has been advanced through time by taking advantage of new principles, tools, techniques, and technology, the novel raised concerns that resulted from multiple challenges complicated the capital investment for infrastructure (Mohammadi et al., 2017). IAM leaders should be aware of the obstacles to prepare organizations for this journey and mitigate risks. The most common challenges to implementing AM for infrastructure are identified in the following subsections.

### *1.4.1 Budget Limitation*

Perhaps budget shrinkage is globally the most common challenge to take care of infrastructure. US infrastructure for many sectors has a huge backlog. For example, $836 billion backlogs of highway and bridge capital needs (i.e., mainly repair and partially expansion-enhancement) (ASCE, 2017). UK local highways backlog has been estimated at £10 billion (Treasury, 2013). This situation brings to matters the need to count on a data-driven decision support model that can dynamically optimize the budget allocation (will be discussed in Chap. 5).

### *1.4.2 Increasing Demand*

The demand for public infrastructure has increased due to fast urbanization and a growing population. American transit systems carried 10.5 billion passenger trips in 2015, which is 33% higher than 20 years ago. Also, an increase in energy consumption of 0.4% per year has been announced from 2015 through 2040 for the energy sector (ASCE, 2017). Increasing demand means a need for upgrading and expanding the infrastructure and imposes additional budget requirements at present for the upgrade or expansion and in the future to preserve and maintain the newly built (and the legacy) assets. This leads to complicating IAM where the impacts of change in demand should be captured in decision-making methodology (Mohammadi et al., 2020) matching with the appropriate upgrade or expansion investments through time-horizon.

### 1.4.3 Aging Infrastructure

In most developed countries, existing infrastructure has been operating since the 1960s and 1970s and in some cases built after World War II. The extensive deterioration of already aged systems complicates managing the network. Therefore, decision-makers need to develop models to simulate the operation and mimic practice. This is the basis for tracking the performance and forecasting future conditions and levels of service for each element necessary to accomplish a safe, reliable, and convenient service. As a tool, the role of deterioration models in the IAM process is to predict the future trend of degradation addressing aging assets (will be discussed in Chap. 3).

### 1.4.4 Higher Society Expectations

CI are social facilities impacting day-to-day public well-being and society economic growth. Expectations to receive higher-quality services and dependency on infrastructure is unlike 20 years or even 10 years ago, while this adds more complexity to managing infrastructure and limits the flexibility.

### 1.4.5 Climate Change

Climate change can't be ignored especially for long-term and forward-thinking strategy while it adds a new criterion in multi-criteria decision making (MCDM) in IAM. For instance, the impacts of future climate in rainfall intensity should be captured in stormwater pipelines (Amador et al., 2020), and increasing the number of freezing cycles needs to be addressed in the pavement maintenance planning (Mohammadi et al., 2019b) (will be discussed in Chap. 6).

### 1.4.6 *Sustainability and Human Development Concerns*

Sustainability is a raised public concern and expectation. Sustainability concerns such as energy efficiency and greenhouse gas (GHG) emissions (Faghih-Imani & Amador Jimenez, 2013) as well as human development goals like poverty alleviation; access to health, education, and fresh water; and gender equity (Mohammadi et al., 2019a) are expected to be addressed in the upcoming plans for public infrastructure (will be discussed in Chap. 6).

### 1.4.7 *Infrastructure Interdependency*

Asset systems within a facility or between different infrastructures are often interdependent. Hence, maintenance of one facility or system often impacts the maintenance activities (or conditions the performance) of other facilities or systems, both economically and functionally. Thus, this interdependency cannot be ignored.

## 1.5 IAM Best Practices

Many governments, municipalities, and institutes recognized the need for developing guidelines and standards to implement IAM. The most common and well-known best practices are:

- *ISO 55000 series,* published by the International Organization for Standardization (ISO) to provide a global language including three guidelines.
  - ISO 55000 (2014a): Asset management—Overview, principles, and terminology.
  - ISO 55001 (2014b): Asset management—Management systems—Requirements.
  - ISO 55002 (2014c): Asset management—Management systems—Guidelines for the application of ISO 55001.
- *International Infrastructure Management Manual (IIMM),* published by the Institute of Public Works Engineering Australasia (IPWEA) (IPWEA, 2020).
- *BSI PAS 55* (IAM and BSI, 2008) *and Asset Management—an Anatomy* (IAM, 2015), published by the British Standards Institution and was initiated by the Institute of Asset Management (IAM).
- *Transportation Asset Management Guide—A Focus on Implementation* (AASHTO, 2013), published by AASHTO.
- *InfraGuide series* (NRC, 2001-2006), published by the Canadian National Research Center (NRC).

All these best practices can be used as benchmarks by organizations seeking to implement AM at an organizational level. However, these references provide only a general understanding of the process, fundamental requirements, and steps to apply AM, while those who are dealing with AM especially for the first time (e.g., small municipalities and agencies, students, and professionals) require detailed and practical steps to fully realize the process and to be able to develop an AM platform from scratch.

This book fills the technical gaps between these guidelines and practical needs by providing and presenting step by step guide collected from the state-of-the-art solutions and already implemented industry best practices.

## 1.6 IAM Main Phases and Maturity

Guidelines like IAM (2015) have developed conceptual AM models which are more or less similar including main components of an organization and people, asset information, decision-making, planning, lifecycle delivery, and risk and review. Each component then will be expanded to multiple subjects. These models are very important to build an AM platform in any organization. For example, how asset management strategy/policy as a leader of the whole process should be drafted or implemented. How the work management system should be designed, maintained, and upgraded toward proactive AM.

As was discussed earlier, this book is aimed to bridge between standards/guidelines and practical needs. For instance, all standards are encouraging for efficient decision-making and being proactive by long-term planning; however, the challenging parts always would be which steps must be taken to achieve this goal, which types of information is required to be collected and prepared, how the decision-making system should be set up, and which tools and techniques can be used to improve decision-making outposts. This book explains those steps, clarifies required information, and presents the most efficient and practical solutions to enhance decision-making and long-term planning for infrastructure systems.

Figure 1.4 presents a simplified process with phases that are required to implement advanced decision-making and long-term planning for IAM. It also generally

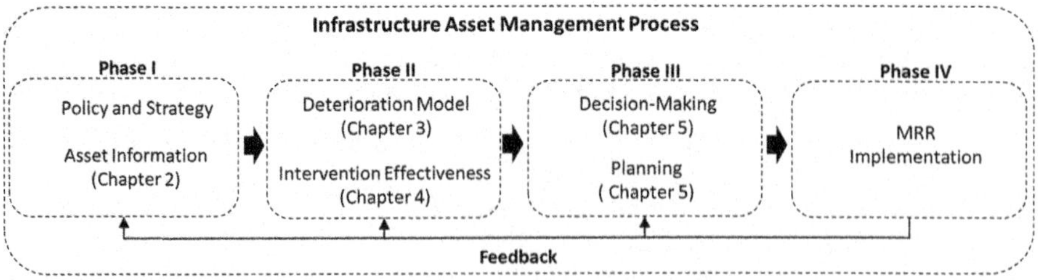


**Fig. 1.4** IAM main phases

presents the scope and layout of this book by linking phases to corresponding chapters.

These phases include preparation (i.e., collecting data and providing inputs for the AM platform), analyzing collected data to provide bases for the next phase (i.e., decision-making), decision-making and planning for asset maintenance, renovation, and replacement (MRR) respecting goals and organization policy, and finally implementing the plan and operation. IAM is a live process and always should be under improvement by addressing feedback from the implementation phase.

These phases cover the required components of IAM; however, each step can be handled in extensive levels of maturities. IAM's level of maturity depends on how the entire organization has been trained and contributes to the process. Senior managers with a good understanding and experience of AM play a key role to support the idea, pushing stakeholders, and providing requirements.

Also, AM maturity is linked to how advanced methods, best practices, tools, and technology are being used and implemented in all steps to achieve the highest return value. The forthcoming chapters provide more details for very basic requirements and less mature platforms as well as more advanced solutions while, obviously, all organizations new to the topic need an evolving time.

## *1.6.1 Phase I: Preparation and Data Collection*

There are different types of required inputs to develop an AM platform for infrastructure.

### 1.6.1.1 Policy and Strategy

At first, the asset owner needs to define principle goals and mandated requirements for this journey. Several factors contribute to how organizations select targets, schedule goals, and prioritize investments. The AM team is responsible to interpret goals (e.g., having safe, comfortable, and reliable subway trains) to measurable SGR targets (e.g., achieving more than 95% reliability in systems/assets or limiting backlogs), and KPIs (e.g., 95% punctuality), around which the whole AM process will be built later. Constraints and limitations (e.g., available budget) also play key roles and must be identified. Best practices like ISO 55000 recommend discussing and developing AM policy at the organization level, which is required to be approved and supported by top managers for the best possible outputs.

#### 1.6.1.2 Asset Information

Asset information is the foundation of IAM, which is always challenging especially at the starting point. Chapter 2 in this book explains what types of data need to be accurately collected, stored, and maintained consistently to build a data inventory for the purpose of IAM.

### *1.6.2 Phase II: Data Analysis*

Collected data must be technically analyzed to extract prediction engines (deterioration models and intervention effectiveness), which are key elements of being proactive and projecting long-term plans.

#### 1.6.2.1 Deterioration Model

To develop cost-effective medium-term and long-term plans, decision-makers need to predict potential future decay in the asset lifecycle. Chapter 3 provides the best practice as well as simplified methods to develop deterioration models.

#### 1.6.2.2 Intervention Effectiveness

Having a real understanding of maintenance effectiveness enhances decision-making and planning to pick more cost-effective solutions (Chap. 4).

### *1.6.3 Phase III: Decision-Making and Planning*

By gathering and analyzing data, the asset management team will be prepared to enter the next phase of planning and decision-making, seeking optimal and cost-effective solutions.

#### 1.6.3.1 Decision-Making

Generally, there are two types of planning during the life cycle of assets: routine and preventive (i.e., proactive) actions, which are typically predefined by manufacturers for electrical and mechanical-based assets or recommended by best practices such as potholes for road pavement and sealing cracks in a concrete building. Another type would be corrective (i.e., reactive) maintenance planning covering major refurbishment, rehabilitation, and replacement actions, commonly named SGR projects. Although, both are part of AM planning, decision-making is more meaningful for corrective, refurbishment, major overhaul, and replacement types of actions, which are financially categorized as capital expenses (CAPEX) compared to operating expenses (OPEX) for routine and preventive maintenance. Decision-making and planning for CAPEX will be discussed in more detail in Chap. 5.

#### 1.6.3.2 Asset Management Plan

The provided interventions for assets should be presented in a practical way to take full advantage of this process by providing an Asset Management Plan (AMP) (Chap. 5).

### *1.6.4 Phase IV: MRR Implementation*

Finally, the planned interventions must be implemented through tactical and operational plans. These cover managing all tasks of implementing preventive and corrective actions, preparing work orders, selecting executors, and asset replacement (i.e., decommissioning and commissioning assets). Typically, organizations take care of routine/preventive maintenance internally and outsource the major corrective works like rehabilitation and replacements. The nature of major rehabilitation and replacement projects is the same as construction/installation projects with all required stages from design to handover.

## 1.7 Exercises

**Exercise 1.1**

Your company has been developing an AM system for the Government of ABC country; so far you have conducted data collection for the entire road network preparing yourself to apply AM for this country. However, in the last Sunday elections, the government lost and there is uncertainty that the new government will abandon the implementation of the Pavement Management Systems (PMS). In an attempt to rescue the project from going into the garbage bin, you have secured a meeting with the new prime minister. Answer the following according to this context.

- Explain what an AM system is; do not use the technical wording; convey a simple message easy to understand by anybody from the public at large, but bear in mind your target is the new prime minister.
- Explain why the AM system is useful and will be used in the future.

**Exercise 1.2**
Identify the main infrastructure systems in your community and explore potential interdependencies.

## References


AASHTO. (2013). *AASHTO transportation asset management guide: A focus on implementation*. U.S. Dept. of Transportation, Federal Highway Administration.

Amador, L., Mohammadi, A., Abu-Samra, S., & Maghsoudi, R. (2020). Resilient storm pipes: a multi-stage decision support system. *Structure and Infrastructure Engineering, 16*(5), 847–859.

American Society of Civil Engineers (ASCE). (2017). 2017 report card for America's infrastructure. http://www.infrastructurereportcard.org. 1 Sept 2020.

American Society of Civil Engineers (ASCE). (2021). The Journal of Infrastructure Systems (Aims & Scope). https://ascelibrary.org/. 1 Nov 2021.

Cambridge Dictionary. (2020). Online dictionary. https://dictionary.cambridge.org/dictionary/English. 1 Sept 2020.

Corporate Finance Institute (CFI). (2020). What is public infrastructure? https://corporatefinanceinstitute.com/resources/knowledge/economics/public-infrastructure/. 1 Sept 2020.

Davis, R. (2012). *An introduction to asset management a simple but informative introduction to the management of physical assets*. The Institute of Asset Management (IAM).

Dawson, R. (2013). Bridges n'that: An infrastructure definition for iBUILD. www.ibuild.ac.uk. 1 Sept 2020.

Elsevier. (2020). International Journal of Critical Infrastructure Protection (IJCIP). https://www.journals.elsevier.com/international-journal-of-critical-infrastructure-protection. 1 Sept 2020.

Faghih-Imani, A., & Amador Jimenez, L. (2013). Towards a sustainable pavement management: Incorporation of environmental impacts of pavement treatments into a performance-based optimization. Transportation Research Record, Journal of the Transportation Research Board of the National Academies.

Federal Highway Administration (FHWA). (2021). Asset management overview. U.S. Department of Transportation. https://www.fhwa.dot.gov. 1 Nov 2021.

Federation of Canadian Municipalities (FCM) and National Research Council (NRC) of Canada. (2005). *The national guide to sustainable municipal infrastructure (InfraGuide): The seventh innovations and best practices, managing infrastructure assets (decision making and investment planning)*. FCM.

FTA. (2010). *Transit asset management practices: A national and international review*. US Department of Transportation.

Fulmer, J. (2009). What in the world is infrastructure? *PEI Infrastructure Investor, 1*(4), 30–32.

Government of Canada. (2020). Critical infrastructure, Canada's Critical Infrastructure. https://www.publicsafety.gc.ca/cnt/ntnl-scrt/crtcl-nfrstrctr/cci-iec-en.aspx. 1 Sept 2020.

Grafius, D. R., Varga, L., & Jude, S. (2020). Infrastructure interdependencies: Opportunities from complexity. *Journal of Infrastructure Systems, 26*(4), 04020036.

Hastings, N. A. J. (2015). *Physical asset management: With an introduction to iso55000 (Second)*. Springer. https://doi.org/10.1007/978-3-319-14777-2

IPWEA. (2020). International infrastructure management manual. 6th Edition. https://www.ipwea.org/publications/ipweabookshop/iimm

Institute for Water Resources (IWR). (2013). Best practices in asset management. https://www.iwr.usace.army.mil/Portals/70/docs/iwrreports/2013-R-08_Best_Practices_in_Asset_Management.pdf

Institute of Asset Management (IAM). (2015). Asset management: An anatomy. https://theiam.org/. 1 Sept 2020.

Institute of Asset Management (IAM) and & British Standards Institution (BSI). (2008). *Asset management: Guidelines for the application of PAS 55-1*. Business Standards institution.

ISO. (2014a). *55000 2014. Asset management. Overview, principles and terminology*. International Organization for Standardization.

ISO. (2014b). *55001 2014. Asset management — Management systems — Requirements*. International Organization for Standardization, Geneva, Switzerland.

ISO. (2014c). *55001 2014. Asset management — Management systems — Guidelines for the application of ISO 55001*. International Organization for Standardization.

Mohammadi, A., Amador-Jimenez, L., & Nasiri, F. (2017). Underground transit system management; new issues. In Canadian transportation research forum 52nd annual conference-Canadian transportation: 150 years of progress//Les transports au Canada: 150 ans de progrès-Winnipeg, Manitoba, 28–31 May 2017.

Mohammadi, A., Elsaid, F., Amador-Jimenez, L., & Nasiri, F. (2019a). Optimizing transit maintenance and rehabilitation to support human development and sustainability: A case study of Costa Rica's railroad network. *International Journal of Sustainable Transportation*, 1–14.

Mohammadi, A., Amador-Jimenez, L., & Elsaid, F. (2019b). Simplified pavement performance modeling with only two-time series observations: A case study of Montreal Island. *Journal of Transportation Engineering, Part B: Pavements, 145*(4), 05019004.

Mohammadi, A., Amador-Jimenez, L., & Nasiri, F. (2020). Reliable, effective, and sustainable urban railways: A model for optimal planning and asset management. *Journal of Construction Engineering and Management, 146*(6), 04020057.

NRC. (2001-2006). *InfraGuide, Nationa Guide to sustainable municipal infrastructure*. National Guide to Sustainable Municipal Infrastructure.

Oxford English and Spanish Dictionary. (2020). online dictionary. https://www.lexico.com/. 1 Sept 2020.

Treasury, H. M. S. (2010). National infrastructure plan 2010. Infrastructure UK, UK Treasury Department (HM Treasury). https://www.gov.uk. 1 Sept 2020.

Treasury, H. M. S. (2013). Investing in Britain's future, UK Treasury Department (HM Treasury). https://www.gov.uk. 1 Sept 2020.

United Nation (UN). (2016) (a). Transforming our world: The 2030 agenda for sustainable development. Retrieved from https://sustainabledevelopment.un.org/. 1 Sept 2020.

United Nations. (1994). *Infrastructure development as key to economic growth and regional economic cooperation*. Economic and Social Commission for Asia and the Pacific, United Nations, New York.

# Chapter 2
# Asset Information

## 2.1 Introduction

This chapter provides an understanding of the required information to develop IAM platforms and the corresponding data management process. AM maturity is highly dependent on the availability and integrity of data. Asset information system feeds the entire AM process and has two main components of asset inventory and work management system.

Asset inventory (also called the asset register) is the critical piece working as a foundation in AM process (Fig. 1.4 and Sects. 2.2, 2.3, and 2.4). Asset inventory starts by identifying assets and collecting data and continues with updating and enhancing information across the lifecycle of assets.

The role of the work management system is to facilitate managing maintenance activities by defining, assigning, tracking, and managing regular work orders for routine maintenance as well as corrective interventions and to maintain records such as frequency and types of actions and associated costs (Sect. 2.5).

## 2.2 Asset Inventory

The development of asset inventory is an initial step in the process of implementing AM in any organization. It serves to identify owned assets, their quantity, location, and current performance/condition. An asset inventory provides a clear picture of the entire system and builds it by adequately registering assets in a structured manner for further utilization. An asset inventory maintains a range of data fields ready to receive data from field and desk studies such as name, location, type, dimension, hierarchy, age, useful life, replacement value (RV), asset criticality, environmental conditions, operational loads, and strength as well as asset current condition. Initial



A. Mohammadi, L. Amador Jimenez, *Asset Management Decision-Making For Infrastructure Systems*, https://doi.org/10.1007/978-3-030-97614-9_2

databases then can be expanded to collect and keep historical records of assessments and interventions for each system, subsystem, and asset.

The comprehensiveness of an asset inventory depends on the organizational policy and experiences in AM and is considered by best practices and guidelines as a key evaluation factor of AM maturity. A basic database at least presents a list of assets, their attributes, and ages while mature platforms provide detailed time-series data covering various aspects and characteristics.

Some organizations may use commercial tools, commonly named enterprise asset management (EAM) systems, to facilitate the process of collecting and maintaining essential data while many others may simply use "flat tables" such as those on a simple Excel spreadsheet lacking relational components and failing to support advanced reporting and visualization over graphical, tabular, or mapping interfaces. These later facilitate work management to define, implement, and track work orders (e.g., routine, preventive, and corrective maintenance) and are commonly called computerized maintenance management systems (CMMS). From an AM perspective, data integrity is the key factor. Therefore, depending on the size and nature of assets, organizations may look at emerging technologies to improve the efficiency and accuracy of the data acquisition process at a fraction of the cost of traditional approaches (APTA, 2013).

Table 2.1 presents a hypothetical dataset for pavement in a road pavement segment. In this example, the required data is divided into four sections of physical attributes, deterioration characteristics, pavement performance, and maintenance history.

The physical attributes provide geographical and identity characteristics of the asset (i.e., road segment) as presented in the example. Physical attributes can also be expanded to provide geographic or projected coordinates, and it can link segments to other features (route events) and other sources of data like geographic information system (GIS) maps and spatial data analysis (Fig. 2.1). This feature would be more important for linear assets such as roads, pipelines, and rail tracks.

Asset utilization characteristics reunite key features that directly serve to predict asset deterioration and serve to identify feasible maintenance alternatives. For example, consider a road segment; the functional (e.g., freeway, highway, arterial, or local roads), structural (e.g., rigid or flexible pavement), traffic load (high, medium, or low demand), and environmental (e.g., wet or dry) features are the main factors. More detail such as the thickness and strength of concrete pavement can also be included. Asset deterioration and cost-effective interventions would be discussed in more detail in Chaps. 3 and 4.

The next section shows how to establish the current level of performance/condition of any asset (e.g., a bridge or road pavement). This is typically one of the most important factors in maintenance decision-making and funding prioritization. In Sect. 2.3, performance/condition assessment, its criticality, and methods are discussed. It is common to encounter that the apparent age of an asset is the best alternative to capture its current level of performance, where there is no access to assessment in relation to its lifespan. All that matters for AM is the ability to establish the current level of performance as contrasted by its ability to deliver

**Table 2.1** Hypothetical dataset for a road pavement segment

| Physical Attributes | | | | | | Deterioration Characteristics | | | | Condition | | Maintenance History | |
|---|---|---|---|---|---|---|---|---|---|---|---|---|---|
| Segment ID | Location | Direction | From Station (Km) | To Station (km) | Lane (No.) | Function | Traffic | Surface | Environmental Condition | Age (year) | IRI (mm/m) | Type | Time |
| H11 | Highway 10 | North | 1.50 | 3.00 | 3 | Highway | High | Concrete | Dry, Non-Freeze | 10 | 2 | Crack Sealing | 2018 |

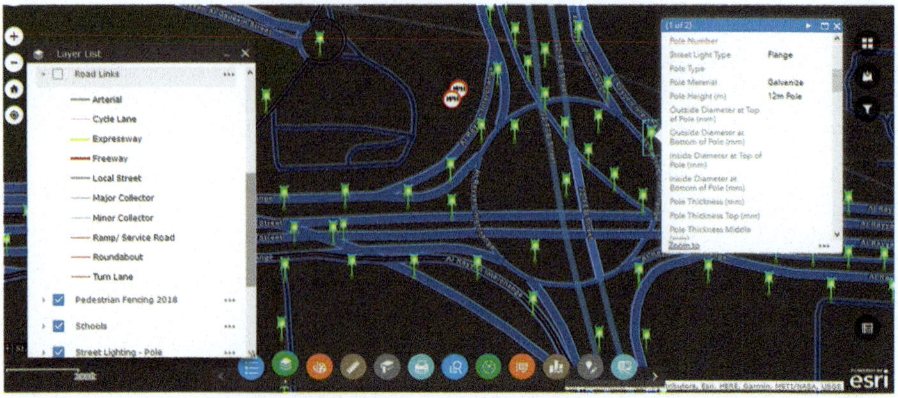


**Fig. 2.1** Sample spatial database for road segments and associated assets [ESRI developer 2020]

public service. An example is presented for one of the most common pavement condition indicators, the international roughness index (IRI), in Sect. 2.3.5.

Finally, maintenance history for each asset or its subcomponents enhances this dataset, which is typically unavailable for many firm owners, especially those who are implementing AM at the early stages.

Similar tables can be developed for other types of assets respecting their nature and functionality. Table 2.2 shows another example, an elevator, in a building.

An elevator is a key component of facility management in the buildings. For example, for electrical and mechanical assets like an elevator, a reliability-based indicator like MTBF (mean time between failure) can be used to present asset performance. The elevator has multi-component, which can be identified as a separate asset while advanced platforms break it down to enhance maintenance planning and apply separate scenarios for each component or even sub-component. This approach may optimize planning and budget allocation, however, increases the costs and complexity of the process. Thus, the optimal level of granularity should be identified by asset managers. Table 2.3 gives the expanded elevator example providing more details for three components. In this example, a quantitative and discrete condition assessment (scaled 1 to 10, higher means better) shows the current condition of assets.

## *2.2.1 System Hierarchy*

Unlike linear systems such as road pavement or railway track, assets in a nonlinear infrastructure (e.g., buildings, stations, and wastewater treatment facilities) are working in systems and subsystems that are linked through a functional hierarchy. Defining hierarchy is a key requirement to manage and maintain assets more

**Table 2.2** Hypothetical Data Inventory for an Elevator

| Physical Attributes | | | Functional Characteristics | | | Condition | | Maintenance History | | Failure History | |
|---|---|---|---|---|---|---|---|---|---|---|---|
| Asset ID | Location | Position | Type | Capacity (Kg) | Manufacturer | Age (year) | MTBF (hours) | Type | Time | Reason | Time |
| MBE1 | Main Building | North | Hydraulic | 1600 | X | 15 | 4000 | Corrective | Sep-17 | Pump | Nov-19 |

**Table 2.3** Expanded Elevator Dataset to Components Level

| Physical Attributes | | | Asset System | | Sub-component | | Performance | | Maintenance History | | Failure History | |
|---|---|---|---|---|---|---|---|---|---|---|---|---|
| Asset ID | Location | Position | Name | Type | Name | Type | Age (year) | Condition | Type | Time | Reason | Time |
| MBE1 | Main Building | North | Elevator | Hydraulic | Pump | Electro motor | 15 | 7 | Corrective | Sep-17 | Motor | Nov-19 |
| MBE1 | Main Building | North | Elevator | Hydraulic | Tank | Metal (200L) | 15 | 6 | Corrective | Oct-17 | – | – |
| MBE1 | Main Building | North | Elevator | Hydraulic | Valve | Control | 5 | 9 | Preventive | Jan-19 | – | – |

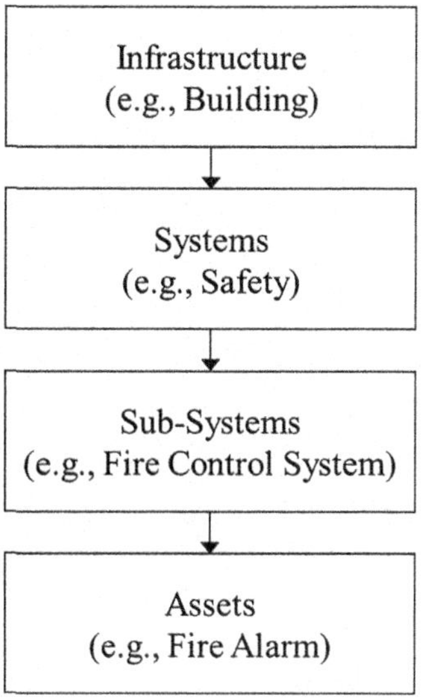


**Fig. 2.2** Asset Functional Hierarchy

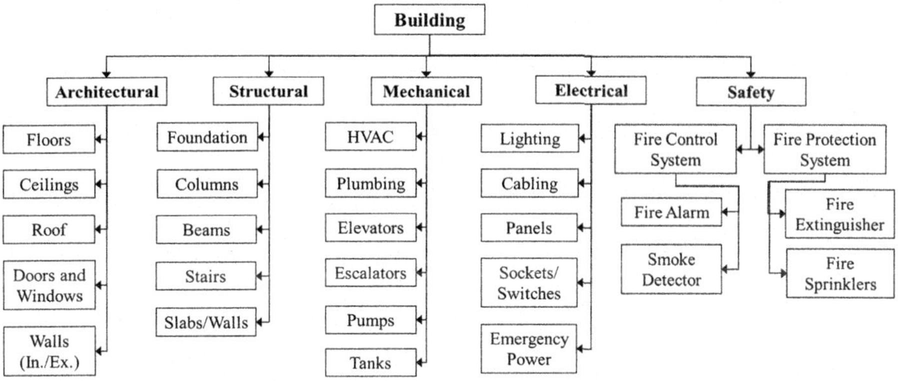


**Figure 2.3** Sample (Building) Functional Hierarchy

efficiently. Thus, in the asset registry process, a proper hierarchy should be identified, and assets would be located in the hierarchy. Figure 2.2 shows a general way of developing the functional hierarchy.

Figure 2.3 presents a sample hierarchy for a building. Five systems of architectural, structural, mechanical, electrical, and safety are identified, while safety is the only system that is assigned subsystems. There are multiple ways to define systems and sub-systems where organizations can reflect current practices and preferences. The number of layers in the hierarchy depends on the complexity of infrastructure, nature of assets, and maintenance scenarios. For instance, in this example, "electrical distribution system" is a potential subsystem for electrical, which covers multiple asset types such as panels and cables. Electrical panels are categorized in single asset types; however, if there are different types of panels with various useful life or preventive and corrective maintenance scenarios, then the proper way would be to use multiple asset classes for panels instead of a single class. There is no right or wrong approach; however, one must make sure that hierarchy covers all physical assets. Then, it should be considered that the way it is structured can impact AM efficiency and applicability.

Organizations that own assets in multiple locations may also develop a location hierarchy. For example, the ministry of education, which is responsible to maintain many schools, can use a location hierarchy to identify the location of assets in each school and corresponding facilities such as the school's main building, play yard, bus loop, etc. to show the location of assets. Depending on how complex an asset portfolio is, a hybrid solution may also be applied where a single hierarchy in the higher levels shows the location of assets and then in sublayers provides functional systems and subsystems.

## 2.3 Performance Assessment

### *2.3.1 System Performance*

The performance of an infrastructure system is evaluated from the quality and availability of provided services. For a water network system, for instance, the reliability and quality of access to clean and fresh water indicate the level of performance. For other systems such as electrical power, telecommunication, and transportation, overall performance can be evaluated similarly. For a transportation system, it is linked to a broad range of criteria such as accessibility, coverage, punctuality, safety, and travel time, while a combination of all these factors represents system performance and alternatively can be called level of service (LOS).

The term LOS has been historically used for transportation systems in the literature to refer to the degree of congestion on a road link, however, recently has been extended to other infrastructure systems in the ample sense of how well the system, or a part of it, delivers its intended (expected) service to the end-user. LOS is nowadays understood as being multidimensional and linked to multiple aspects of the system (and its assets) including the reliability of the service, the safety to users' health, the convenience, the cost, and in a nut-shell all aspects that matter to the end-user as explained in the following sections. More and more, the literature refers to Key Performance Indicators (KPIs) when assessing the LOS in the specific context of a given system.

Upkeeping and maintaining acceptable performance at the system level is the main goal in the AM process for infrastructure. Balzer and Schorn (2015) introduced three categories, namely, technical, sociological, and economical for infrastructure system performance. Conditions, reliability, and efficiency indicate the technical performance of the system while consumers' perspectives and levels of satisfaction can be captured by sociological indicators. The business level in terms of cost and revenue shows how economically the system is working. Traditionally, asset

**Table 2.4** KPIs for Urban Railway Systems

| Category | Area | KPI |
|---|---|---|
| Technical | Functionality | Service Quality |
| | | Service Efficiency |
| | Reliability | Failure Rate |
| | | Percent Available |
| | Sustainability | Energy Consumption |
| | | Gas Emission |
| | Coverage | Connectivity |
| | | Accessibility |
| | Safety | Accident Rate |
| | | Accident Severity |
| Sociological | Travel Time | Punctuality |
| | | Journey Time |
| | Travel Cost | Ticket Price |
| | | Lost Hours |
| | Information | Time/Cost Option |
| | | Real-Time |
| | Security | Vulnerability |
| | | Violence Risk |
| | Comfort and Safety | Thermal/ Vibration |
| | | Noise/Lighting |
| Economical | Revenue | Ticket |
| | | Advertisement |
| | Cost | OPEX |
| | | CAPEX |

managers mostly concentrate on technical metrics to run decision-making for managing assets. Engaging the stakeholders has been recently recommended by several regulatory bodies through their guidelines and standards, such as ISO 55000 (2014) and Water Services Regulation Authority in England and Wales (OFWAT) as the main requirement for future asset management (Jones et al., 2014), which is less invested in the current practices (Karam, 2016). However, infrastructure is often publicly managed by non-profit organizations and companies where economic indicators are not ranked first and thus give less attention.

The organization's corporate goals are the main source to extract KPIs (in the technical, sociological, and economical dimensions) and to compare services against them; such assessment later guides decision-making and budget allocation. Selecting KPIs depends on several factors including stakeholders' preferences; access to equipment, technology, and trained operators; available budget; and predefined goals and strategy by politicians. Table 2.4 shows the hierarchical expansion of KPIs for urban railway systems (Mohammadi et al., 2018).

Similar KPIs can be identified for a water supply network and Table 2.5 is one example of such KPI for this type of infrastructure.

**Table 2.5** KPIs for Water Supply Network

| Category | Area | KPI |
|---|---|---|
| Technical | Reliability | Failure Rate |
| | | Percent Available |
| | Sustainability | Energy Consumption |
| | | Gas Emission |
| | Coverage | Percentage of Area |
| | Safety | Pollution |
| Sociological | Quality | Color |
| | | Smell |
| | Cost | Unit Rate |
| | | Connection |
| | Information | Text/Email |
| | | Real-Time |
| Economical | Revenue | Bill |
| | Cost | OPEX |
| | | CAPEX |

### 2.3.2 *Asset Performance*

Asset performance, on the other hand, indicates the performance of each asset individually as well as a part of a system. System performance and overall LOS for infrastructure come from integrating each and every subsystem/asset performance back into the system. For instance, many subsystems and assets contribute to rail transit on-time performance (i.e., punctuality) where a failure or drop of performance, of a single asset, in the rail track (e.g., switch), in the rail car (e.g., braking system), in the control system (e.g., signaling), or even in a station (e.g., platform) may cause a delay in the entire network. It would be necessary that all subsystems and assets in a system work properly to provide adequate services; however, it doesn't guarantee to result in the user expected level of services unless all subsystems and assets work well as a system.

Punctuality can be also influenced by other internal and external factors. Delay in a rail corridor "A" can cause a delay in corridor "B", or human error in scheduling in the main control center of the network may cause a delay in this corridor, which is not necessarily an asset failure in corridor "B". It can also result from external reasons such as breaking a water pipeline in a region and closing the underground tunnel or rail corridor.

Although several internal and external factors contribute to system KPIs, clearly all subsystems and assets must provide an acceptable performance as a primary requirement. Hence, from an AM perspective, the ultimate goal is to maintain and upkeep LOS and performance at the system level; however, it is required to focus first on individual subsystem and asset performance linking to the overall KPIs. Therefore, in practice, maintenance, refurbishment, rehabilitation, and replacement for infrastructure are planned at a subsystem/asset level.

As part of the asset inventory and a starting point, the current (i.e., at the time of decision-making) performance of each asset should be determined. Asset performance investigates the current state of the asset, which helps decision-makers to identify critical elements, prioritize interventions, and invest in the right place. Assets naturally lose performance and reliability over time and under operation; however, premature failures can also happen. The main objective of measuring and assessing asset performance is addressing the loss of functionality, finding elements in a critical situation, and optimizing investment.

### 2.3.3 *Performance Versus Condition*

Evaluating the physical condition of assets to determine asset state, gears, and decay often results in asset "condition"; however, performance represents the level of provided services by asset rather than physical condition. There is a mutual relation between performance and condition of assets while organizations sometimes use these terms interchangeably to present the current state. However, performance-based assessment is recognized as a more mature method compared to condition-based due to addressing service metrics. Performance-based planning and decision-making are more challenging especially for civil infrastructure such as roads, bridges, and water pipelines to prioritize assets based on the quality of provided services. For this reason, it became the most common practice to use the physical condition of the asset instead of performance to decide and plan in IAM.

However, as was discussed earlier, maintenance intervention planning should be linked to system-level performance and user concerns. It means that the classical AM paradigm of condition-based (i.e., asset-based and asset-oriented) should be enhanced in near future to become service based and user oriented. Chap. 6 will provide some example methodologies to address this gap in current industry practice.

A traditional condition indicator, which may present the current state, is the age of the asset. This is still common among many organizations to use the age of assets to plan for the future. This age can be an apparent age referring to the time between either the first building (or installation) or a major rehabilitation and the current time. However, the deterioration trend of assets can be accelerated or decelerated due to many sorts of environmental, physical, and operational factors, that are not necessarily captured by age of assets. Therefore, the age of assets should not be the sole indicator to judge the maintenance and replacement time for assets. For some assets such as a street luminaire lamp (which typically works until failure and no maintenance or intervention is feasible), asset age can be the only condition indicator, while for most civil, mechanical, and even electrical assets, one or more condition indicators can be devised.

For organizations, the solution usually would not be as simple as doing assessments where condition measurement is often a time-consuming and very expensive process, which is not frequently affordable by many asset owners. Therefore, small organizations need to find a cost-effective and balanced strategy in the type and frequency of assessments. The optimal scenario depends on considering several factors including the nature and type of assets; the scale and quantity of assets; criticality and consequence of failures; age and probability of failures, decay slope; and the deterioration model. A trustable deterioration model (will be discussed in more detail in the next chapter) helps decision-makers to avoid frequent assessments saving time and money.

### 2.3.4 *Qualitative Versus Quantitative Assessment*

Assets may be evaluated qualitatively or quantitatively. Qualitative assessment, mostly based on expert judgment, has been used for a long time to evaluate assets. Although this approach has been reviewed and upgraded by defining guidelines and grading procedures to avoid subjective assessments, however, still has several limitations that prevent its use on mature IAM. Qualitative assessment limits decision-making to ranking or worst-first approaches and does not allow for mathematical optimization, which will be discussed in more detail later in Chap. 5. Also, having a qualitative assessment in a hierarchical system challenges the aggregation of assets and subsystems to provide a system's overall performance. Besides, using a limited number of predefined discrete levels (e.g., excellent, good, or fair) limits the flexibility of evaluation for proper prioritization where there are thousands of assets and many of them are graded at the same levels. Table 2.6 shows a sample qualitative assessment for multi-element assets such as a bridge, which is adapted from (ASCE, 2017).

A quantitative assessment gives a chance for preparing numerical scales and sometimes more precise evaluation, which then can be used for various types of planning and decision-making including mathematical optimization. Depending on the nature of indicators, different ranges can be used. A classical quantitative approach is applying a 1 to 5 or 1 to 10 scale. The advantage of this method is

**Table 2.6** A Sample Qualitative Assessment Guideline

| | |
|---|---|
| A: Exceptional | A Brand new or recently rehabilitated asset.<br>A few elements may show signs of decay. |
| B: Good | A few elements deteriorated significantly.<br>Some elements require attention due to signs of deterioration. |
| C: Mediocre | General signs of deterioration and requires attention.<br>Some elements show significant defects and loss of functionality. |
| D: Poor | Asset in a below-standard condition while many elements approaching the end of their service life.<br>The majority of the element shows significant deterioration. |
| F: Failing | An unacceptable condition for the asset while many elements show signs of failure. |

Adapted from (ASCE, 2017)

**Table 2.7** A Sample of Classical Quantitative Performance Assessment

| Qualitative | Exceptional | Good | Mediocre | Poor | Failing |
|---|---|---|---|---|---|
| Quantitative | 5 | 4 | 3 | 2 | 1 |

applicability to multi-discipline infrastructure such as buildings and rail transits where there are different types of assets like electrical, mechanical, and civil assets, and this solution enables assessing the entire system with a common language. It partially improves qualitative method gaps, however, still provides only a discrete assessment. Table 2.7 provides a sample numerical rating system that can be used to evaluate asset conditions using the Table 2.6 guideline.

Another main limitation of these approaches is being subjective and highly dependent on inspectors' experience. Therefore, researchers and practitioners have invested time and effort to improve such processes making them less human dependent and more technology based. This resulted in more modern assessments, which are not subjective and provide continuous numerical indicators, such as IRI for road pavements. These kinds of assessments involve advanced equipment and technology, like sensors, to capture more accurately the current state of assets.

Qualitative ranks are useful in the higher levels of reporting where non-technical policymakers and politicians are looking for a bigger picture of current and future conditions. Infrastructure report cards (e.g., ASCE, 2017; CIRC, 2016) are becoming more common among governments to observe nation infrastructure and provide an assessment by qualitative indicators. However, in a mature approach, numerical indexes are used in the lower levels of hierarchy, and then the overall conditions are presented for the public by qualitative levels.

## *2.3.5 Condition-Based Assessment Methods*

In this section, condition assessment methods (both classical and advanced) are presented for selected main types of infrastructure.

### 2.3.5.1 Pavement

Road pavement is the pioneer in the AM practices and most advanced methodologies have been applied first on pavement, possibly due to less complexity of components compared to a building, transit system, or even water network. One direction that shows this maturity is access to a variety of advanced condition indicators:

**IRI** International Roughness Index, or IRI, was developed by the World Bank in the 1980s (Sayers et al., 1986) and converts the pavement wheel path profile using a mathematical model to pavement ride quality indices. This indicator is also

**Table 2.8** Typical IRI Thresholds

| Levels | Exceptional | Good | Fair | Poor |
|---|---|---|---|---|
| IRI (m/km) | < 1.0 | 1–1.66 | 1.67–2.37 | 2.37< |

**Table 2.9** Typical PCI triggers (Adjusted from Ontario Ministry of Transportation (2013))

| | | Road Function | | | |
|---|---|---|---|---|---|
| | PCI | Freeway | Arterial | Collector | Local |
| Condition Level | Good | 75 | 70 | 65 | 60 |
| | Fair | 74–66 | 69–56 | 64–51 | 59–46 |
| | Poor | 65 | 55 | 50 | 45 |

addressed in ASTM standards of ASTM E1926 – 08 (ASTM, 2015) and ASTM E1364 – 95 (ASTM, 2017). This widely used indicator presents road smoothness per m/km and inch/mile, while smaller IRI means smoother pavement.

A brand-new highway is supposed to achieve a 0.5–0.6 m/km while other road functions like arterial or local get higher IRI due to the type of structure and lower smoothness requirement. However, the material and construction quality, as well as agency requirements (i.e., design), may also impact road smoothness. IRI increases as the road experiences weather cycles and traffic loads. Pavement performances drop from "Good" to "Fair" and then "Poor" levels. Entering in a poor boundary often means the end of the service life for pavements and a need for replacement. Best practices provide thresholds for acceptable and unacceptable smoothness levels and triggers for ending service life. Chen et al. (2019) reviewed 241 research participants' perspectives on IRI thresholds for flexible roads. Table 2.8 shows the results of this study; however, agencies may adjust these thresholds depending on their policy.

**PCI** Pavement condition index (PCI) has been proposed by ASTM D6433–20 (ASTM 2020). This standard provides a guideline to identify and measure the severity, quantity, and density of different types of distresses. Then PCI will be estimated by deducting distresses from 100 (i.e., a brand-new road) as can be seen in Eq. 2.1.

$$\text{PCI} = 100 - \sum \text{Deducts (distresses)} \tag{2.1}$$

Unlike IRI, which mostly reflects the surface condition of roads and not necessarily the structural performance and defects, PCI considers several types of distresses in the pavement including its lower structural layers. Similar boundaries as Table 2.8 can be developed for PCI. Table 2.9 presents PCI condition triggers for different road functions, which is adjusted from the Ontario Ministry of Transportation (2013) guideline for road design and rehabilitation.

There are more pavement condition metrics such as VIR (visual inspection rating), SAI (structural adequacy index), SDI (surface distress index), and PCR (pavement condition rating), which have their methodology; however, PCI and IRI are the most common indicators among road agencies.

Selecting the right indicator could be a challenge especially for small agencies with less flexibility to pick several options. Access to equipment, machinery, and qualified teams as well as corresponding costs to assess can be the primary criteria. PCI and IRI became the most common pavement indicators across the world, and the best standards and guidelines defined thresholds and intervention triggers for these two indicators. However, still, these two methods can be compared technically to pick the right one.

IRI advantages

- Directly reflects customer concerns by addressing pavement roughness
- There are low-cost methods of assessment for this indicator

IRI disadvantages

- Mostly reflects surface distresses
- May miss small cracks. Thus, it is more of a lagging indicator (Tan, 2015).

PCI advantages

- Easy to understand due to presenting on a 0–100 scale especially for non-technical audiences.
- Addresses both surface and structure defects
- More trustable for preventive maintenance as can be a leading indicator

PCI disadvantages

- Since PCI integrates both structure and surface defects, it may confuse the asset manager in selecting a proper action.

Therefore, selecting either IRI or PCI may not address all concerns. For instance, dropping (i.e., increasing) IRI of a road segment can be caused by surface defects while issues related to under layers also may result in road roughness. Planning maintenance only based on the IRI assessment may mislead selecting the right intervention. Improving only surface conditions where there are structural defects means wasting money and time. Thus, the ideal solution would be assessing the pavement condition using more than one indicator addressing both structural and surface conditions. For example, a combination of IRI and PCI or IRI and SAI provides a better underhanging for the nature of stresses leading to optimal planning. This is the reason both agencies and condition assessment consultants are recently approaching to provide more comprehensive assessments.

**Table 2.10** FHWA Condition Assessment System

| Quantitative Index | Qualitative Index | Description |
|---|---|---|
| 0 | Failed | Out of service |
| 1 | Failure | Major deterioration |
| 2 | Critical | Advanced deterioration |
| 3 | Serious | Loss of section, deterioration, spalling, or scour |
| 4 | Poor | Advanced section loss, deterioration, spalling, or scour |
| 5 | Fair | May have minor section loss, cracking, spalling, or scour |
| 6 | Satisfactory | Minor deterioration |
| 7 | Good | Some minor problems |
| 8 | Very good | No problem noted |
| 9 | Excellent | – |

Adapted from FHWA (1995)

#### 2.3.5.2 Bridge

From a safety perspective, condition assessment in a bridge is much more critical than pavement. Failure in the pavement can cause unsafe situations for users; however, failure in a bridge often results in social, economic losses with fatalities, major injuries, and subsequent road closure. Bridge inspection guidelines have been developed to evaluate the condition and reliability of a bridge and its main components including the deck, superstructure, substructure, culvert, and channel. However, after several collapses, those guidelines become more and more expanded in order to avoid subjective assessment by inspectors.

Guidelines now define specific procedures for selecting the right inspector with adequate knowledge and experience providing rules for the quality and frequency of inspections needed. The National Bridge Inspection Standards (NBIS) for the United States published by FHWA is one of those standards. A rating system of 0 to 9 (9 means excellent) is proposed in this standard. Table 2.10 presents a summarized and adapted version of the condition rating system defined by FHWA (1995).

FHWA-HRT-15-081 (Chase et al., 2016) reviewed the state of the art in the United States and other countries respecting bridge condition indices including the bridge health index (BHI) and bridge condition index (BCI). It highlighted that one main limitation of current condition assessment is relying on a visual inspection and subjective assessments. Similar approaches have been used by other countries like Canada, the United Kingdom, Japan, and Germany. In an experimental study in the United States, 49 inspectors from 25 states were asked to inspect several bridges, and results indicated that 68% of primary element condition ratings for individual bridge components were varied within one point of the average while 95% varied within two rating points (Phares et al., 2004).

Therefore, improving traditional condition assessment methods for bridges is always interested by scholars (Ghodoosi et al., 2016). There are different types of non-destructive tests (NDTs) to inspect bridges such as ultrasonic or ground-penetrating radar (GRP) (Ryan et al. 2012).

**Structural Health Monitoring (SHM)**
A more advanced technique of bridge assessment is structural health monitoring (SHM) (Zanini et al., 2019). Sensors are installed on bridge elements to observe its condition and health by collecting acceleration, deflection, and other distress. The recent advances in sensors and wireless technology allow for real-time observations. This data then would be proceeded to evaluate the bridge condition (Shahsavari et al., 2020). This technique provides continuous observations of this critical asset to avoid failures as well as unnecessarily maintenance (FPrmeC Solutions, 2019).

#### 2.3.5.3 Water and Wastewater Network

Water distribution networks, as well as storm and sanitary systems, are also the major infrastructure for all urban areas. Similar methods like Table 2.10 can be used to implement a quantitative assessment for these types of assets; however, the assessment would not be easy as pavement or even bridges because most of the network is underground with often no access for the exterior side while darkness and weather conditions make also very difficult inner side inspection. Furthermore, for most pipes due to size, there is no chance of manual inspection. This resulted in more popularity of age-based maintenance planning for this type of asset.

**Acoustic-Based Assessment**
An alternative solution for assessing buried pipelines is an acoustic wave test to collect health information. Existing pipes are attached acoustic sensors to induce a sound wave capturing the time it takes to travel between two sensors. Collected data then is analyzed (comparing original and current thickness) to assign a grade (e.g., good [5 and 4], moderate [3], or poor [2 and 1]) and to determine the average minimum wall thickness. The acoustic signal can also detect leaks (SUEZ, 2021).

**Image-Based Assessment**
Engineers also tried to take advantage of technology and develop more advanced performance assessment techniques using laser profiler, sewer electro scan, sonar, zoom camera, and digital scanning. Using a robot and camera mounted on top of a crawler or a float became more popular recently, named closed circuit television (CCTV) technique (Kaddoura, 2015). The camera provides a bunch of 2-D or 3-D photos and videos, which in the next step must be analyzed to identify defects. This is typically handled by certified operators through inspecting videos, which is time consuming, labor intensive, and error prone. Implementing image processing using artificial intelligence (AI) techniques such as deep learning to automate this process can facilitate, accelerate, and improve the accuracy of performance assessment (Moradi et al., 2020). Identified defects and distresses must be interpreted to performance indicators and ranks for further actions (Kaddoura et al., 2018).

#### 2.3.5.4 Building

Key building envelopes such as hospitals, schools, universities, and courts are examples of multi-discipline infrastructure systems where naturally different assets such as structural, electrical, and mechanical components work together as a system to provide expected services for stakeholders. Other examples of similar infrastructure are transit systems and power plants.

Several technical indicators can be defined to assess performance in the building such as availability (e.g., elevator), comfort (e.g., thermal), and reliability (e.g., electrical power). Newly raised concerns include energy-saving and GHG emissions that can address the level of sustainability. Due to the criticality of building envelopes, several standards and guidelines are proposed to improve and secure performance in the building (ASCE, 2000; ASTM, 2000; ASTM, 2015; and ISO 11863, 2011).

A numerical discrete condition assessment system (e.g., a 1–5 scale) has been commonly used for this infrastructure. This approach can be applied to all types of assets and provides a uniform assessment output.

**Facility Condition Index (FCI)**

One of the most common indicators for evaluating building conditions is the facility condition index (FCI). The FCI is calculated by dividing the existing cost of deferred maintenance (i.e., corrective and replacement) by the current RV. To estimate this indicator, inspectors first decide about current and near-future (usually 3–5 years face time) required CAPEX investments. This comes from reviewing failure history as well as current condition, age, or other performance indicators (KPIs). Then, FCI is calculated by comparing the required investment with the current RV (Eq. 2.2). Although it has a monetary indicator, FCI provides a quantitative measure of asset condition, stated as a percentage.

$$\text{FCI} = \frac{\text{Total Deferred CAPEX (\$)}}{\text{RV (\$)}} \times 100 \tag{2.2}$$

The higher the percentage, the poorer the condition of the building. To interpret the FCI depending on the selected facetime, the criticality of building and organization policy ranges should be defined. This is one sample set of thresholds:

- 0–10%, Good
- 10–30%, Fair
- 30–100%, Poor

Meanwhile, several limitations can be identified for FCI:

- This condition indicator could be more applicable to evaluate the whole system (e.g., a building) rather than subsystems and assets.
- Could not address safety in critical assets, subsystems, and systems. Higher FCI means more need for repair but may not be attributable to a health or safety issue.
- Less chance for long-term trade-off planning.

- Does not necessarily reflect the physical condition of assets.
- Impossible to directly address and focus on other concerns such as LOS and sustainability.

**Example 2.1**
Table 2.11 summarizes the current and future required CAPEX and RV for a building in the coming 10 years. Assuming there is no capital budget to spend and estimate FCI for the first 5 years considering 3 years face time.

***Solution*** The FCI for the first year would be estimated as below:

$$\mathrm{FCI}_1 = \frac{25,000 + 10,000 + 5,000}{1,000,000} \times 100 = 4\%$$

Since there is no capital budget, the cumulative required budget of years 1–4 will be used to calculate FCI for year 2 while year 1 is commonly known as backlog (Table 2.12). This example shows ignoring investment at the right time how fastly can push buildings to poor conditions.

Comprehensive platforms are also proposed to capture more features in building condition assessments. Galasiu et al. (2019) reviewed more than 200 references for assessing building functional suitability and identified these two platforms as the most comprehensive frameworks to assess real property assets:

- The ASTM Standards on Whole Building Functionality and Serviceability (WBFS) (2000) as an internationally recognized buildings' life cycle AM methodology, which includes 19 individual standards covering over 100 topics of building serviceability and 340 building features, each with levels of service calibrated from 0 (not present, does not have, not applicable). Also, two standards published by ISO (ISO 11863, 2011; ISO 15686-10, 2010) incorporate the ASTM WBFS methodology.
- The BUILDER Sustainment Management System (SMS), which is a web-based software, provides AM service by prioritizing building components using a knowledge-based inspection (KBI) methodology to generate building condition index (BCI) and functionality index (FI). This platform was developed by the US Army Corps of Engineering Research and Development Centre, Construction Engineering Research Laboratory (ERDC-CERL) (2012) to enhance maintenance practices for public building infrastructure.

### 2.3.6 Performance-Based Assessment Methods

As was discussed earlier, it is not common to use performance-based and service-based indicators for civil infrastructure assets like a road, bridge, or pipeline. This partially comes from the nature of these assets and the type of service they provide.

**Table 2.11** Dataset

| Year | 1 | 2 | 3 | 4 | 5 | 6 | 7 | 8 |
|---|---|---|---|---|---|---|---|---|
| Required CAPEX ($) | 25,000 | 10,000 | 5000 | 45,000 | 15,000 | 18,000 | 100,000 | 25,000 |
| RV ($) | 1,000,000 | 1,020,000 | 1,040,400 | 1,061,208 | 1,082,432 | 1,104,081 | 1,126,162 | 1,148,686 |

**Table 2.12** Solution

| Year | 1 | 2 | 3 | 4 | 5 | 6 | 7 | 8 |
|---|---|---|---|---|---|---|---|---|
| Required CAPEX ($) | 25,000 | 10,000 | 5000 | 45,000 | 15,000 | 18,000 | 100,000 | 25,000 |
| RV ($) | 1,000,000 | 1,020,000 | 1,040,400 | 1,061,208 | 1,082,432 | 1,104,081 | 1,126,162 | 1,148,686 |
| FCI (%) | 4.00 | 8.33 | 9.61 | 11.12 | 20.14 | | | |

However, multi-discipline systems such as transits, buildings, and water systems are recommended to take advantage of a performance-based approach for decision-making and budget distributions. In this section, sample performance-based assessment methods from industry best practices are provided.

#### 2.3.6.1 Reliability-Based Assessment

For non-structural assets and equipment such as mechanical (e.g., pump and elevator), electrical (e.g., lighting and control panel), and safety (e.g., fire alarm), still, a numerical condition-based indicator such as a 1–5 scale can be implemented to assess the condition of the assets. However, reliability-based indicators (adapted from the manufacturing industry) are also very common, especially among multi-discipline infrastructure such as energy plants, buildings, and transportation systems.

These indicators are categorized as performance-based methods representing the services rather than physical condition. Also, compared to non-analytic and subjective approaches (e.g., 1–5 scale), reliability-based indicators are data-driven, which improve significantly the quality and accuracy of assessment as well as provide non-discrete outputs. By collecting and analyzing data; reviewing failure history; and benchmarking from best practices, these performance indicators can be estimated. Here, the main reliability KPIs to measure performance are reviewed (Fiix, 2020).

**Mean Time to Failure (MTTF)**
This metric indicates the time that a non-repairable asset lasts until a failure happens (e.g., a lightbulb). This metric shows the remaining service life for the asset, which often is provided by manufacturers or can be updated by users. It is always key to be ready ahead before failure in the system by replacing assets and providing extra equipment or spare parts.

**Mean Time Between Failure (MTBF)**
This similar indicator represents the average time between failure (one failure to another), while is more applicable for a system or multi-component asset in which one or more components or sub-component passed MTTF. The time between breakdowns is crucial from a performance measurement perspective. Mean Distance Between Failure (MDBF) is the alternative metric for bus and rail fleet types systems (Eq. 2.3).

$$\text{MTBF} = \frac{\text{Total Operation Time}}{\text{Number of Failures}} \tag{2.3}$$

**Mean Time To Repair (MTTR)**
This indicator shows how fast is the maintenance process and it measures the time between when the failure happened and the system returning to the normal operation (Eq. 2.4).

$$\text{MTTR} = \frac{\text{Total Time Spent on Maintenace}}{\text{Number of Failures}} \tag{2.4}$$

**Availability**

This is another key indicator in this topic, which represents the percentage of time that asset or system of assets is performing properly by comparing the uptime (i.e., normal performance time) and downtime (i.e., the asset is not available, which can be planned or unplanned) (Eq. 2.5).

$$\text{Availability} = \frac{\text{Uptime (MTBF)}}{(\text{Uptime (MTBF)} + \text{Downtime (MTTR)})} \tag{2.5}$$

**System Reliability**

Finally, system reliability provides the probability that a system can work in normal performance without failure. The literature is quite rich in this topic and several formulations are proposed to calculate reliability (Dhillon, 2004; Birolini, 2013), but this is one simplified example (exponential) formulation to estimate reliability based on MTBF or MTTF (Eq. 2.6).

$$\text{Reliability } (t) = e^{-\gamma t} \qquad \gamma = \frac{1}{\text{MTBF or MTTF}} \tag{2.6}$$

All these metrics can be used to plan maintenance interventions and replacements. However, one should keep in mind that the nature of failure for electrical and mechanical assets is often different from civil infrastructure assets. For instance, any failure in a bridge (even one component or sub-component like a beam or column) may result in a disaster, and measuring MTBF may not be feasible. For a pavement, it is completely different where we never experience a failure with a meaning that is covered by these metrics. A road pavement may lose performance and can show many distresses and potholes, which means losing safety, comfort, and service but still the road is not totally gone.

Another key limitation of this type of assessment for civil infrastructure is less chance of projecting the future in long-term planning. The nature of these indicators especially in the asset level is more short-term, showing current operation status rather than being predictable to develop a deterioration model, for instance. Therefore, implementing a reliability-based approach in AM planning may not be applicable for all types of assets and may not help to apply all types of analyses. Some transit organizations tried to collect massive historical data and develop a degradation model for fleet MDBF per age of the fleet; however, this still is at the fleet level, not system or asset levels.

### 2.3.7 Overall Condition

Enhancing the overall condition of a system is always one main objective in IAM, while to run quantitative decision-making, it is required to be able to estimate the overall condition, dynamically observing the impact of any budgeting scenario in services. For instance, the condition of a specific highway is assessed by dividing it into some segments and the overall condition comes from integrating the condition of all segments while using a simple average would not be a smart approach.

For linear infrastructure such as road pavements or pipelines, a simple additive weighting (SAW) approach (Eq. 2.7) can be a better alternative to estimate the overall condition. For example, for road *m*, which has *I* segments ($S_{i,m}$, I: 1 to I), the overall condition of the road, $P_{m,}$ can be calculated by this equation while $w_i$ represents segment weight (Eq. 2.8). The length of the segment or surface (where the number of the lane is changing) is commonly used for this purpose. Traffic load is also an alternative when a road agency prefers to guide budget allocation according to the number of users by implementing a similar formulation like Eq. 2.8.

$$P_m = \sum_{i=1}^{I} w_i S_{i,m} \tag{2.7}$$

$$w_i = \frac{l_i}{\sum_{i=1}^{I} l_i} \quad \text{while} \quad \sum_{i=1}^{I} w_i = 1 \tag{2.8}$$

There would be a rationale behind dividing the highway and defining segments, which will be discussed in the next chapter.

**Example 2.2**
A road agency recently inspected 20 segments of a highway using IRI, and Table 2.13 shows the inspected results. What would be the overall IRI for this 65-kilometer highway?

***Solution*** Based on Eq. 2.8, each segment weight can be calculated using segment length, and then by adding up weighted IRI (Eq. 2.7), the overall IRI for this highway would be estimated at 1.30 (m/km). If the length of segments is ignored, the average IRI for this highway changes to 1.41(m/km) indicating the role of weights (Table 2.14).

For a non-linear system such as a building or a subway system, while different types of assets are linked in a hierarchical network, it would be more complex to estimate the overall condition. Figure 2.4 presents one way to define a hierarchy for building components and sub-components. Depending on the level of accuracy and criticality that the IAM team requires, this can be expanded to lower levels to include elements or even sub-elements.

**Table 2.13** A Sample Dataset for Example 2.2

| Segment No. | Segment Length (km) | Segment IRI (m/km) |
|---|---|---|
| 1 | 2.30 | 0.80 |
| 2 | 3.00 | 0.95 |
| 3 | 2.20 | 1.70 |
| 4 | 3.50 | 1.10 |
| 5 | 3.50 | 2.68 |
| 6 | 1.90 | 1.30 |
| 7 | 4.20 | 0.65 |
| 8 | 3.60 | 0.60 |
| 9 | 2.50 | 1.00 |
| 10 | 1.50 | 2.45 |
| 11 | 3.20 | 2.20 |
| 12 | 4.00 | 2.00 |
| 13 | 3.90 | 1.20 |
| 14 | 1.75 | 1.70 |
| 15 | 3.40 | 0.90 |
| 16 | 3.35 | 1.55 |
| 17 | 3.65 | 1.40 |
| 18 | 2.85 | 1.35 |
| 19 | 2.45 | 0.85 |
| 20 | 4.00 | 1.90 |

To estimate the overall building condition, first the condition of the lowest level (asset or subsystem such as stairs and pumps in this example) must be assessed. Then, all assets/subsystems conditions are aggregated to estimate the system's overall condition. In the next step, the overall condition of the building can be calculated by combining systems conditions.

Equation 2.7 can also be utilized for each level in this system; however, since assets are naturally different, a challenge would be defining an adjusted weighting approach. One solution would be using RV to estimate weights, but this approach does not reflect the criticality of each asset, subsystem, and system. Generally, for such nonlinear infrastructure, it would be more complex to find the role and impact of increasing or dropping the condition of each component in the overall condition.

MCDM problems are very close to these cases where different alternatives are compared based on several criteria (i.e., asset component here) and sub-criteria (asset sub-components here) while similarly each criterion or sub-criterion may have various weight in estimating the total scores of each alternative, which could be our infrastructure system (e.g., building). For instance, the bidding process for a construction project is an MCDM problem, and general contractors are compared based on several criteria such as previous experience, financial support, and machinery and the winner is who gets the highest scores.

AHP (Saaty, 1980) is a very common method to estimate weights in MCDM problems. This is a pairwise comparison method based on expert judgments and surveys. Experienced people will be asked to compare two by two asset

**Table 2.14** Example 2.2 solution

| Segment No. | Segment Length (km) | Segment IRI (m/km) | Segment Weight | Weighted IRI |
|---|---|---|---|---|
| 1 | 2.30 | 0.80 | 0.0354 | 0.0283 |
| 2 | 3.00 | 0.95 | 0.0462 | 0.0438 |
| 3 | 2.20 | 1.70 | 0.0338 | 0.0575 |
| 4 | 3.50 | 1.10 | 0.0538 | 0.0592 |
| 5 | 3.50 | 2.68 | 0.0538 | 0.1443 |
| 6 | 1.90 | 1.30 | 0.0292 | 0.0380 |
| 7 | 4.20 | 0.65 | 0.0646 | 0.0420 |
| 8 | 3.60 | 0.60 | 0.0554 | 0.0332 |
| 9 | 2.50 | 1.00 | 0.0385 | 0.0385 |
| 10 | 1.50 | 2.45 | 0.0231 | 0.0565 |
| 11 | 3.20 | 2.20 | 0.0492 | 0.1083 |
| 12 | 4.00 | 2.00 | 0.0615 | 0.1231 |
| 13 | 3.90 | 1.20 | 0.0600 | 0.0720 |
| 14 | 1.75 | 1.70 | 0.0269 | 0.0458 |
| 15 | 3.40 | 0.90 | 0.0523 | 0.0471 |
| 16 | 3.35 | 1.55 | 0.0515 | 0.0799 |
| 17 | 3.65 | 1.40 | 0.0562 | 0.0786 |
| 18 | 2.85 | 1.35 | 0.0438 | 0.0592 |
| 19 | 2.45 | 0.85 | 0.0377 | 0.0320 |
| 20 | 4.00 | 1.90 | 0.0615 | 0.1169 |
| | | Overall IRI (m/km) | | **1.30** |

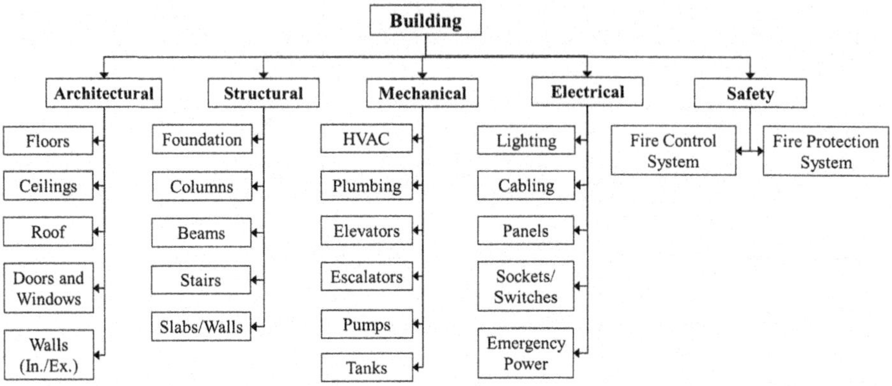


**Fig. 2.4** Sample way to define building asset hierarchy

sub-components from a criticality perspective and their role in the component level condition. A similar approach can be applied at the asset component level to estimate each component's weight and importance for the overall condition of the building.

AHP method assumes that there is no interdependency between asset sub-components and asset components while it does not necessarily reflect reality for many systems such as a building. For instance, the performance of the several mechanical components is highly dependent on the electrical components' reliability meaning that there is interdependency between machinal and electrical assets. ANP (Saaty, 2001) is an alternative method for such systems, which follows a similar methodology to AHP, with more pairwise comparisons and complexity in the calculation. Both AHP and ANP models can be developed on the Microsoft Excel spreadsheet; however, ANP would be more complex. There are non-commercial decision-maker software that can be used to run these MCDM methods and other alternatives.

In the meantime, using SAW formulation may result in misleading for a hierarchical system where an overall outcome of a system with a combination of very high-condition and very low-condition assets can be presented at a good and acceptable level. This can lead to missing sub-components in critical conditions. Alternative MCDM methods such as technique of order preference similarity to the ideal solution (TOPSIS) (Hwang & Yoon, 1981) or multi-attribute utility theory (MAUT) try to partially address this gap. A weighted geometric average representing the MAUT method (Eq. 2.9) is one example that can be used instead of SAW (Eq. 2.7) to recover this limitation. In this formulation, the overall condition in the upper level of hierarchy ($P_m$) is estimated by aggregating the lower level asset condition ($A_{i,\ m}$) where $I$ represents the number of assets in the lower level.

$$P_m = \left[\prod_{i=1}^{I} (A_{i,m})^{w_i}\right]^{1/\sum_{i=1}^{I} w_i} \tag{2.9}$$

**Example 2.3**

The mechanical system in the building (Fig. 2.4) has been recently inspected applying a 1–5 scale (5 means excellent condition). The criticality of each sub-component is also determined by experts as presented in Table 2.15. What would be the overall condition of the mechanical system in this building using both SAW and MAUT?

***Solution*** SAW method

$$P_m = \sum_{i=1}^{I} w_i S_{i,m} = 5 * 0.2 + 2 * 0.05 + 1 * 0.30 + 3 * 0.25 + 4 * 0.10 + 5 * 0.10 = 3.05$$

MAUT method

$$P_m = \left[\prod_{i=1}^{I} (A_{i,m})^{w_i}\right]^{1/\sum_{i=1}^{I} w_i}$$

**Table 2.15** Example 2.3 dataset

| | HVAC | Plumbing | Elevators | Escalators | Pumps | Tanks |
|---|---|---|---|---|---|---|
| Condition (1–5) | 5 | 2 | 1 | 3 | 4 | 5 |
| Weight (%) | 20 | 5 | 30 | 25 | 10 | 10 |

$$P_m = \left(5^{0.20} \times 2^{0.05} \times 1^{0.30} \times 3^{0.25} \times 45^{0.10} \times 5^{0.10}\right)^{\frac{1}{\sum(0.20+0.05+0.30+0.25+0.10+0.10)}} = 2.53$$

Comparing the final results for these two methods shows the role of changing the formulation.

## 2.4 Data Collection

Leveraging data acquisitions and developing asset inventory is a time-consuming and expensive process especially for underdeveloping AM platforms. This would be a system collaboration where different parts of the organization contribute to providing a complete database. Firm owners may outsource and hire consulting companies to gather such data and populate the asset inventory. Also, as was mentioned earlier, the platform for storing data can be commercial software with extra features such as access to cloud or smartphone apps. Mature systems use a single platform to maintain data and manage maintenance works.

Collecting data and developing an asset inventory should be a periodic process (APTA, 2013). Frequent updates are necessary, and more advanced systems will implement site instrumentation to feed in real-time observations over IT systems (i.e., Weigh-in-motion, Bluetooth sensors, LiDAR technologies, etc.). Larger organizations in developed countries are heading to use the advantages of data technology such as cloud data collecting and storing information in a real-time environment. However, many smaller organizations still have the challenge to provide the basic required data to ensure the AM platform can work properly and accurately.

Asset owner needs to define a procedure to collect required data frequently to support the capability of making the right decision at the right time to achieve the organizational goals (i.e., achieve high asset value across the system). Typically, the development of the initial dataset requires the definition of a data structure through relational databases (i.e., data structured in tables and relations to support queries, reporting, and visualization). Collecting data from scratch and registering all assets for the first time would be more challenging, time-consuming, and expensive while organizations should look for longer terms considering future needs and goals.

Section 2.2 provided different types of data that should be collected for a mature asset inventory. Except for the condition assessment process, the rest of the data collection process includes fewer complex and technical steps of identifying, locating, and categorizing assets; counting and measuring assets; gathering current and historical information; and finally processing data into the database. Organizations need to define procedures to capture data from newly built and installed assets as well as to update any changes that happened during the asset lifecycle. The key consideration would be managing well the procedure to ensure that accurate, accessible, comprehensive, consistent, and real-time asset inventory is created and maintained.

Condition assessment, however, is a more technical and challenging step, needs to define inspection and evaluation procedure, and may need specific equipment and facilities. Common 1–5 assessment (Table 2.7) is sometimes only visual, which requires a trained team and clear guidelines to avoid subjective evaluation, however, still is often the only option for many assets.

Taking advantage of technology, new tools can be utilized to improve the quality and speed of assessment avoiding subjective evaluations. For instance, Automatic Road Analyzers (ARAN) vehicles provide a full pavement inspection with several indicators such as IRI, rutting, and laser inspection for cracks. Figure 2.5 shows a sample ARAN, and road owners need to just drive this car on the planned road at the required speed!

Figure 2.6 presents an example of using advanced technology for data acquisition in pipelines for image-based assessment (Sect. 2.3.5.3).

Depending on the type and size of the asset inventory, collected data can proceed into Microsoft Access or Excel spreadsheets. Alternatively, asset owners may use commercial software for this purpose to store and maintain data; however, unless dealing with a huge dataset (Big Data), Excel spreadsheets would be a very low-cost alternative particularly for local organizations.

### *2.4.1 Low-Cost Technique for Condition Assessment*

Using technology has multiple advantages like reducing human error and accelerating the process. Also, for some assets like underground and small pipes, there is less

**Fig. 2.5** A sample ARAN (Cafiso et al., 2019)

**Fig. 2.6** Robot for pipeline inspection (Ciszewski et al., 2020)

**Fig. 2.7** The axis accelerometer board in smartphones (Su & Twu, 2020)

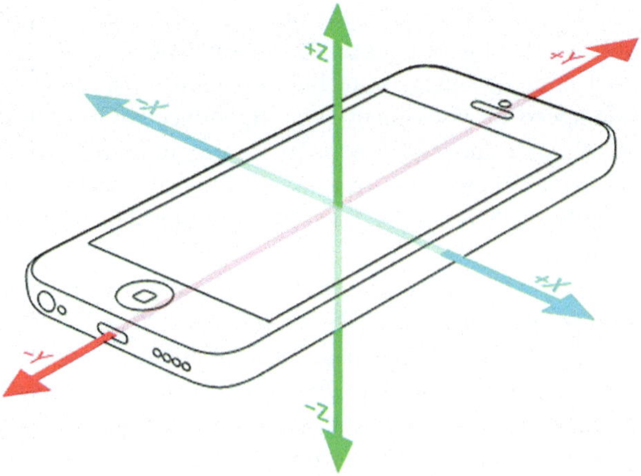


chance for visual inspection by a technician. However, technology is sometimes very expensive for many asset owners like small municipalities. Thus, researchers tried to take advantage of low-cost tools (e.g., smartphones) to provide affordable options for all users. Smartphones have several capabilities, for instance, the accelerometer embedded in new phones captures vertical accelerations, correlated to road surface conditions. Accelerometer serves many motion-related applications, captures accelerations in three dimensions (X, Y, and Z) including lateral and longitudinal accelerations (Fig. 2.7).

Several free and commercial apps can be used for data acquisition (Fig. 2.8). Most apps are able to provide records in CSV and Excel formats.

Amador-Jimenez and Matout (2014) showed that the normalized longitudinal speed by the standard deviation of vertical acceleration provides an estimate for IRI. Eqs. 2.10, 2.11, and 2.12 show how measured acceleration can be interpreted to IRI where

$a_{zi}$: z-acceleration
$a$: mean of the z-acceleration set
$N$: number of z-accelerations considered
$v_{yi}$: speed (velocity) in the y-direction

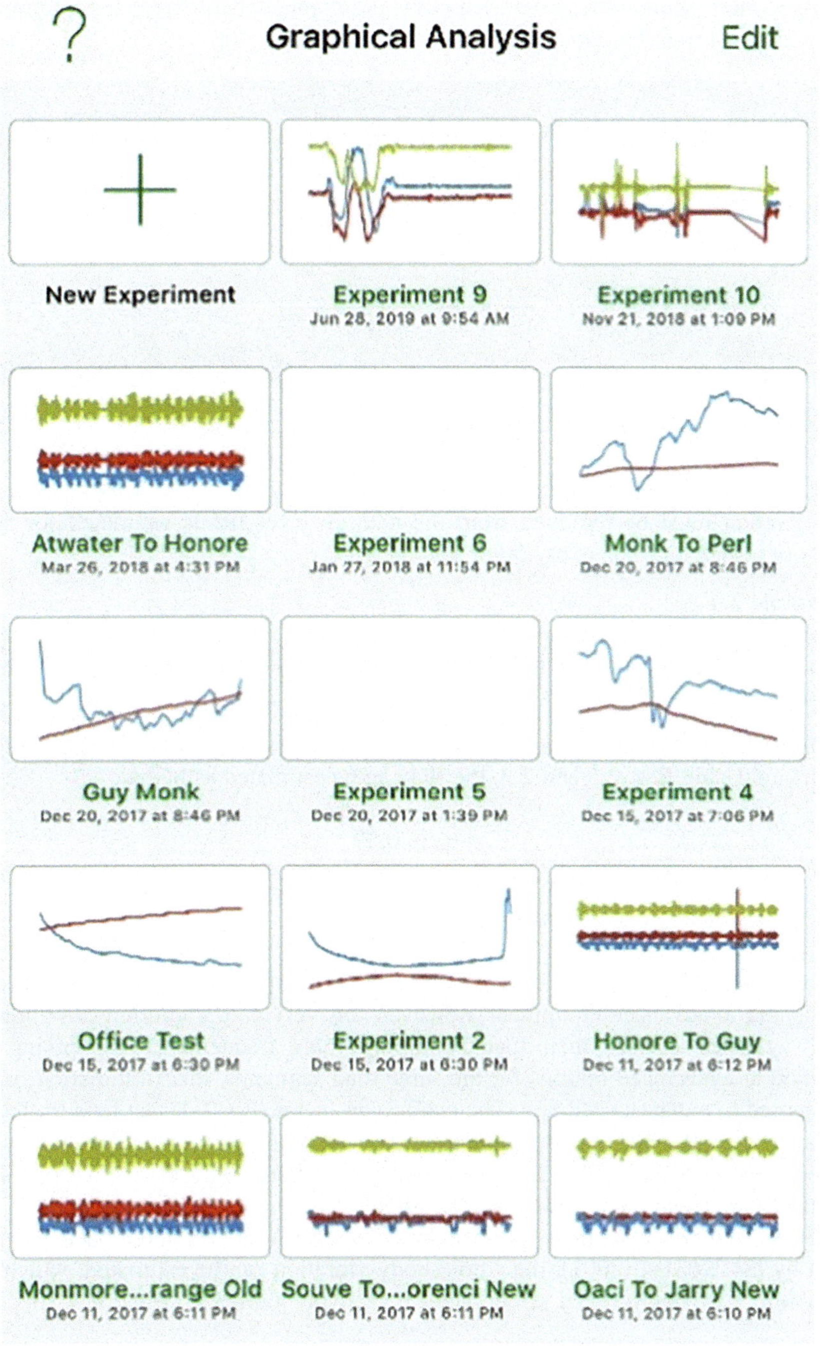


**Fig. 2.8** One sample app for collecting acceleration (Vanier Graphical Analysis)

$\beta$: Correction factor, 100 can be used but is preferable to calibrate using a sample of roads with trustable IRI values.

$$\sigma_z = \sqrt{\frac{1}{n}\sum_{i=1}^{N}(\partial_{zi} - \partial)^2} \tag{2.10}$$

$$\frac{\sigma_z}{v_{yi}} = \frac{\sqrt{\frac{1}{n}\sum_{i=1}^{N}(\partial_{zi} - \partial)^2}}{v_{yi}} \tag{2.11}$$

$$\text{Roughness Indicator (IRI)} = \frac{\sigma_z}{v_{yi}} \times \beta \tag{2.12}$$

The smartphone must be fixed on a car moving at a constant speed (70–80 km/h can be a good option for highways). Therefore, longitudinal acceleration and deceleration must be removed from the data. Typical IRI is estimated for small parts of a segment (i.e., every 50 or 100 meters).

**Example 2.4**
Table 2.16 provides a sample collected vertical acceleration and longitudinal speed (m/s) by a smartphone mounted in a car on a highway. What would be the highway IRI (m/km)?

***Solution*** This dataset shows observations for 2 seconds (around 50 meters) of this highway. Using Eqs. 2.2 and 2.3, the IRI can be estimated as below:

$$\sigma_z = 0.29$$

$$\text{IRI}\left(\frac{\text{m}}{\text{km}}\right) = \frac{\sigma_z}{v_y} = \frac{0.29}{24.957} \times 100 = 1.18$$

Nobody should expect to achieve the same level of accuracy for this approach compared to advanced equipment; however, this very cheap solution can allow all road agencies to track their road conditions more frequently. By applying this method and advanced options for the same road segments, this formulation can be calibrated to enhance accuracy. The calibration factor (Eq. 2.4) can be adjusted for this purpose. Recently, smarter cellphone apps are proposed to the customers, which directly estimate IRI and even provide support to estimating PCI while can automatically map and color-code roads (OtalPave, 2020).

Also, by collecting three-dimensional acceleration and recommended formulation by ISO 2631-4 (2001), the whole-body vibration can be estimated, which is a very important comfort and safety factor for railway journeys. Equation 2.13 shows

**Table 2.16** Sample Collected Acceleration and Speed for a Road

| Time (s) | Vertical Acceleration (m/s$^2$) | Longitudinal Speed (m/s) |
|---|---|---|
| 0.00 | 9.3392 | 24.957 |
| 0.10 | 9.3392 | 24.957 |
| 0.20 | 9.4962 | 24.957 |
| 0.30 | 10.281 | 24.957 |
| 0.40 | 10.281 | 24.957 |
| 0.50 | 10.281 | 24.957 |
| 0.60 | 9.4177 | 24.957 |
| 0.70 | 9.4962 | 24.957 |
| 0.80 | 9.4962 | 24.957 |
| 0.90 | 9.8101 | 24.957 |
| 1.00 | 9.8101 | 24.957 |
| 1.10 | 9.8101 | 24.957 |
| 1.20 | 9.967 | 24.957 |
| 1.30 | 9.7316 | 24.957 |
| 1.40 | 9.5746 | 24.957 |
| 1.50 | 9.5746 | 24.957 |
| 1.60 | 9.6531 | 24.957 |
| 1.70 | 9.967 | 24.957 |
| 1.80 | 9.967 | 24.957 |
| 1.90 | 9.8101 | 24.957 |
| 2.00 | 9.6531 | 24.957 |

how measured three-dimensional acceleration in a railway transit can be converted to whole-body acceleration, where $\alpha_{wx}$, $\alpha_{wy}$, and $\alpha_{wz}$ are the frequency-weighted *root mean square* acceleration in (m/s$^2$) and $k_x$, $k_y$, and $k_z$ are multiplying factors. Then calculated accelerations are compared to recommended thresholds for safety and comfort (Mohammadi et al., 2020).

$$a_v = \sqrt[2]{(k_x\, a_{wx})^2 + \left(k_y\, a_{wy}\right)^2 + \left(k_z\, a_{wz}\right)^2} \tag{2.13}$$

Using supplementary sensors (e.g., Node+) (Variableinc, 2017), it would be possible to collect and transfer more features such as thermal (humidity and temperature), air quality ($CO_2$), and noise. Mohammadi et al. (2020) proposed a methodology to evaluate railway car passengers' comfort as a KPI for AM of rail transit systems by using smartphone technology (Fig. 2.9).

**Fig. 2.9** Supplementary devices to capture more features by smartphone (Variableinc, 2017)

### 2.4.2 *Big Data*

Technology allows collecting real-time observations, using vehicle or infrastructure sensors, and this means dealing with big data. Applying sensors technology and recording data in the cloud, a huge amount of data is being collected per second and even milliseconds. Such data must be filtered, cleaned, classified, processed, and then can be used for AM purposes, while collecting and maintaining big-size data would not be easy and may not necessarily result in enhancing the process. Therefore, the optimum level of detail, types, and frequency of data recording should be optimized. Recently, working with big data became facilitated thanks to AI and data mining techniques such as machine learning and deep learning. Therefore, depending on the type, size, and benefit of collected data, IAM may need to implement AI techniques to prepare data for further steps.

## 2.5 Exercises

**Exercise 2.1**
Which pavement indicator provides both surface and under layers conditions?

A. IRI
B. PCI
C. SAI
D. None

**Exercise 2.2**
One company decided to evaluate the condition of the road. They have a mix of recently rehabilitated as well as some old segments with different defects in under layers. Which performance indicator do you recommend?

A. Using IRI
B. Using PCI
C. Using PCI and IRI
D. Using IRI and VIR (Visual Inspection Report)

## Exercise 2.3

A municipality recently hired Mike. He used his cellphone and collected accelerations on a highway to have an idea about pavement conditions. The below table shows his collected data which provides vertical acceleration and longitudinal speed. Can you help him to do this assessment? What would be the overall condition? What do you think about this road? (Good, Fair, or Poor?) He prefers to have IRI for segments of 50 meters in length (Table 2.17).

**Table 2.17** Collected Data by Cellphone for Road Journey

| Time | Acce. Z (m/s2) | Long. Speed (m/s) | Time | Acce. Z (m/s2) | Long. Speed (m/s) | Time | Acce. Z (m/s2) | Long. Speed (m/s) |
|---|---|---|---|---|---|---|---|---|
| 46:44.8 | 9.1038 | 22.82 | 46:46.4 | 9.7316 | 24.47 | 46:48.0 | 9.967 | 24.47 |
| 46:44.8 | 9.1038 | 23.05 | 46:46.4 | 9.7316 | 24.47 | 46:48.0 | 10.0455 | 24.47 |
| 46:44.8 | 9.4962 | 23.28 | 46:46.4 | 9.6531 | 24.47 | 46:48.0 | 10.0455 | 24.47 |
| 46:44.9 | 9.7316 | 23.52 | 46:46.4 | 9.6531 | 24.47 | 46:48.0 | 9.6531 | 24.47 |
| 46:44.9 | 9.7316 | 23.75 | 46:46.4 | 9.7316 | 24.47 | 46:48.0 | 9.8101 | 24.47 |
| 46:44.9 | 10.124 | 23.99 | 46:46.5 | 9.6531 | 24.47 | 46:48.0 | 9.8101 | 24.47 |
| 46:44.9 | 9.5746 | 24.23 | 46:46.5 | 9.6531 | 24.47 | 46:48.1 | 9.8886 | 24.47 |
| 46:44.9 | 9.5746 | 24.47 | 46:46.5 | 9.8101 | 24.47 | 46:48.1 | 9.6531 | 24.47 |
| 46:45.0 | 9.8886 | 24.47 | 46:46.5 | 9.7316 | 24.47 | 46:48.1 | 9.967 | 24.47 |
| 46:45.0 | 9.8886 | 24.47 | 46:46.5 | 9.6531 | 24.47 | 46:48.1 | 9.967 | 24.47 |
| 46:45.0 | 10.3594 | 24.47 | 46:46.6 | 9.6531 | 24.47 | 46:48.1 | 9.8886 | 24.47 |
| 46:45.0 | 9.5746 | 24.47 | 46:46.6 | 9.8101 | 24.47 | 46:48.2 | 10.0455 | 24.47 |
| 46:45.0 | 9.3392 | 24.47 | 46:46.6 | 9.6531 | 24.47 | 46:48.2 | 10.0455 | 24.47 |
| 46:45.1 | 9.3392 | 24.47 | 46:46.6 | 9.6531 | 24.47 | 46:48.2 | 9.5746 | 24.47 |
| 46:45.1 | 9.4962 | 24.47 | 46:46.6 | 9.8101 | 24.47 | 46:48.2 | 9.6531 | 24.47 |
| 46:45.1 | 10.281 | 24.47 | 46:46.7 | 9.7316 | 24.47 | 46:48.2 | 9.6531 | 24.47 |
| 46:45.1 | 10.281 | 24.47 | 46:46.7 | 9.8101 | 24.47 | 46:48.3 | 9.6531 | 24.47 |
| 46:45.1 | 10.281 | 24.47 | 46:46.7 | 9.8101 | 24.47 | 46:48.3 | 9.8886 | 24.47 |
| 46:45.2 | 9.4177 | 24.47 | 46:46.7 | 9.6531 | 24.47 | 46:48.3 | 9.8886 | 24.47 |
| 46:45.2 | 9.4962 | 24.47 | 46:46.8 | 9.7316 | 24.47 | 46:48.3 | 9.8886 | 24.47 |
| 46:45.2 | 9.4962 | 24.47 | 46:46.8 | 9.7316 | 24.47 | 46:48.4 | 10.0455 | 24.47 |
| 46:45.2 | 9.8101 | 24.47 | 46:46.8 | 9.8886 | 24.47 | 46:48.4 | 9.6531 | 24.47 |
| 46:45.2 | 9.8101 | 24.47 | 46:46.8 | 9.6531 | 24.47 | 46:48.4 | 9.6531 | 24.47 |
| 46:45.2 | 9.8101 | 24.47 | 46:46.8 | 9.6531 | 24.47 | 46:48.4 | 9.5746 | 24.47 |
| 46:45.3 | 9.967 | 24.47 | 46:46.8 | 9.6531 | 24.47 | 46:48.4 | 9.7316 | 24.47 |
| 46:45.3 | 9.7316 | 24.47 | 46:46.9 | 9.7316 | 24.47 | 46:48.4 | 9.7316 | 24.47 |
| 46:45.3 | 9.5746 | 24.47 | 46:46.9 | 9.967 | 24.47 | 46:48.5 | 9.8886 | 24.47 |
| 46:45.3 | 9.5746 | 24.47 | 46:46.9 | 9.967 | 24.47 | 46:48.5 | 9.8886 | 24.47 |
| 46:45.3 | 9.6531 | 24.47 | 46:46.9 | 9.8886 | 24.47 | 46:48.5 | 9.8886 | 24.47 |
| 46:45.4 | 9.967 | 24.47 | 46:46.9 | 9.8101 | 24.47 | 46:48.5 | 9.7316 | 24.47 |
| 46:45.4 | 9.967 | 24.47 | 46:47.0 | 10.124 | 24.47 | 46:48.5 | 9.7316 | 24.47 |
| 46:45.4 | 9.8101 | 24.47 | 46:47.0 | 10.124 | 24.47 | 46:48.6 | 9.7316 | 24.47 |

(continued)

**Table 2.17** (continued)

| Time | Acce. Z (m/s2) | Long. Speed (m/s) | Time | Acce. Z (m/s2) | Long. Speed (m/s) | Time | Acce. Z (m/s2) | Long. Speed (m/s) |
|---|---|---|---|---|---|---|---|---|
| 46:45.4 | 9.6531 | 24.47 | 46:47.0 | 9.6531 | 24.47 | 46:48.6 | 9.6531 | 24.47 |
| 46:45.4 | 9.7316 | 24.47 | 46:47.0 | 9.3392 | 24.47 | 46:48.6 | 9.6531 | 24.47 |
| 46:45.5 | 9.7316 | 24.47 | 46:47.0 | 9.3392 | 24.47 | 46:48.6 | 9.7316 | 24.47 |
| 46:45.5 | 10.124 | 24.47 | 46:47.1 | 9.8886 | 24.47 | 46:48.6 | 9.7316 | 24.47 |
| 46:45.5 | 9.6531 | 24.47 | 46:47.1 | 9.8101 | 24.47 | 46:48.7 | 10.0455 | 24.47 |
| 46:45.5 | 9.6531 | 24.47 | 46:47.1 | 9.8101 | 24.47 | 46:48.7 | 10.0455 | 24.47 |
| 46:45.6 | 9.3392 | 24.47 | 46:47.1 | 9.8101 | 24.47 | 46:48.7 | 10.0455 | 24.47 |
| 46:45.6 | 9.8886 | 24.47 | 46:47.1 | 9.8886 | 24.47 | 46:48.7 | 9.8886 | 24.47 |
| 46:45.6 | 10.124 | 24.47 | 46:47.2 | 9.8886 | 24.47 | 46:48.8 | 9.2607 | 24.47 |
| 46:45.6 | 10.124 | 24.47 | 46:47.2 | 9.8886 | 24.47 | 46:48.8 | 9.8101 | 24.47 |
| 46:45.6 | 9.5746 | 24.47 | 46:47.2 | 9.6531 | 24.47 | 46:48.8 | 9.8101 | 24.47 |
| 46:45.6 | 9.967 | 24.47 | 46:47.2 | 9.5746 | 24.47 | 46:48.8 | 9.967 | 24.47 |
| 46:45.7 | 9.967 | 24.47 | 46:47.2 | 9.967 | 24.47 | 46:48.8 | 9.7316 | 24.47 |
| 46:45.7 | 9.8886 | 24.47 | 46:47.3 | 9.967 | 24.47 | 46:48.8 | 9.7316 | 24.47 |
| 46:45.7 | 9.5746 | 24.47 | 46:47.3 | 9.8101 | 24.47 | 46:48.9 | 9.5746 | 24.47 |
| 46:45.7 | 9.8886 | 24.47 | 46:47.3 | 9.6531 | 24.47 | 46:48.9 | 9.967 | 24.47 |
| 46:45.7 | 9.8886 | 24.47 | 46:47.3 | 9.6531 | 24.47 | 46:48.9 | 9.8886 | 24.47 |
| 46:45.8 | 9.7316 | 24.47 | 46:47.3 | 9.7316 | 24.47 | 46:48.9 | 9.8886 | 24.47 |
| 46:45.8 | 9.967 | 24.47 | 46:47.4 | 9.8886 | 24.47 | 46:48.9 | 9.5746 | 24.47 |
| 46:45.8 | 9.967 | 24.47 | 46:47.4 | 9.8101 | 24.47 | 46:48.9 | 9.5746 | 24.47 |
| 46:45.8 | 9.8886 | 24.47 | 46:47.4 | 9.8101 | 24.47 | 46:49.0 | 9.5746 | 23.50 |
| 46:45.8 | 9.6531 | 24.47 | 46:47.4 | 9.967 | 24.47 | 46:49.0 | 9.8886 | 23.30 |
| 46:45.9 | 9.4962 | 24.47 | 46:47.4 | 9.8101 | 24.47 | 46:49.0 | 9.5746 | 23.00 |
| 46:45.9 | 9.4962 | 24.47 | 46:47.5 | 9.8101 | 24.47 | 46:49.0 | 9.7316 | 22.80 |
| 46:45.9 | 9.7316 | 24.47 | 46:47.5 | 9.7316 | 24.47 | 46:49.1 | 9.7316 | 22.70 |
| 46:45.9 | 9.5746 | 24.47 | 46:47.5 | 9.7316 | 24.47 | 46:49.1 | 9.967 | 22.50 |
| 46:46.0 | 9.5746 | 24.47 | 46:47.5 | 9.8101 | 24.47 | 46:49.1 | 9.8101 | 22.00 |
| 46:46.0 | 9.4177 | 24.47 | 46:47.6 | 9.8101 | 24.47 | | | |
| 46:46.0 | 9.5746 | 24.47 | 46:47.6 | 9.8101 | 24.47 | | | |
| 46:46.0 | 9.8101 | 24.47 | 46:47.6 | 9.8886 | 24.47 | | | |
| 46:46.0 | 9.8101 | 24.47 | 46:47.6 | 9.8886 | 24.47 | | | |
| 46:46.0 | 9.6531 | 24.47 | 46:47.6 | 9.8886 | 24.47 | | | |
| 46:46.1 | 9.8886 | 24.47 | 46:47.6 | 9.5746 | 24.47 | | | |
| 46:46.1 | 9.8886 | 24.47 | 46:47.7 | 9.6531 | 24.47 | | | |
| 46:46.1 | 9.8886 | 24.47 | 46:47.7 | 9.6531 | 24.47 | | | |
| 46:46.1 | 9.6531 | 24.47 | 46:47.7 | 9.967 | 24.47 | | | |
| 46:46.1 | 9.3392 | 24.47 | 46:47.7 | 10.0455 | 24.47 | | | |
| 46:46.1 | 9.3392 | 24.47 | 46:47.7 | 10.0455 | 24.47 | | | |
| 46:46.2 | 9.6531 | 24.47 | 46:47.7 | 9.6531 | 24.47 | | | |
| 46:46.2 | 9.8886 | 24.47 | 46:47.8 | 9.6531 | 24.47 | | | |
| 46:46.2 | 9.8886 | 24.47 | 46:47.8 | 10.0455 | 24.47 | | | |

(continued)

**Table 2.17** (continued)

| Time | Acce. Z (m/s2) | Long. Speed (m/s) | Time | Acce. Z (m/s2) | Long. Speed (m/s) | Time | Acce. Z (m/s2) | Long. Speed (m/s) |
|---|---|---|---|---|---|---|---|---|
| 46:46.2 | 10.0455 | 24.47 | 46:47.8 | 10.0455 | 24.47 | | | |
| 46:46.2 | 9.8886 | 24.47 | 46:47.8 | 9.4962 | 24.47 | | | |
| 46:46.3 | 9.5746 | 24.47 | 46:47.8 | 9.8101 | 24.47 | | | |
| 46:46.3 | 9.5746 | 24.47 | 46:47.9 | 9.8101 | 24.47 | | | |
| 46:46.3 | 9.8101 | 24.47 | 46:47.9 | 9.967 | 24.47 | | | |
| 46:46.3 | 9.7316 | 24.47 | 46:47.9 | 9.967 | 24.47 | | | |
| 46:46.4 | 9.7316 | 24.47 | 46:47.9 | 9.6531 | 24.47 | | | |

**Table 2.18** Exercise 2.4 Database

| Segment No. | Segment Length (km) | Number of Lane | Traffic (AADT) | Segment PCI |
|---|---|---|---|---|
| 1 | 2.30 | 2 | 25,000 | 85.00 |
| 2 | 3.00 | 2 | 30,000 | 88.00 |
| 3 | 2.20 | 2 | 35,000 | 90.00 |
| 4 | 3.50 | 2 | 54,000 | 78.00 |
| 5 | 3.50 | 2 | 55,000 | 6.00 |
| 6 | 1.90 | 2 | 80,000 | 45.00 |
| 7 | 4.20 | 2 | 89,000 | 55.00 |
| 8 | 3.60 | 2 | 35,000 | 30.00 |
| 9 | 2.50 | 3 | 24,000 | 38.00 |
| 10 | 1.50 | 3 | 55,000 | 83.00 |
| 11 | 3.20 | 3 | 80,000 | 44.00 |
| 12 | 4.00 | 3 | 89,000 | 40.00 |
| 13 | 3.90 | 3 | 35,000 | 75.00 |
| 14 | 1.75 | 3 | 32,000 | 95.00 |
| 15 | 3.40 | 3 | 48,000 | 45.00 |
| 16 | 3.35 | 2 | 43,000 | 55.00 |
| 17 | 3.65 | 2 | 79,000 | 30.00 |
| 18 | 2.85 | 2 | 75,000 | 38.00 |
| 19 | 2.45 | 2 | 60,000 | 88.00 |
| 20 | 4.00 | 2 | 46,000 | 90.00 |

### Exercise 2.4

The below table summarizes PCI inspections for a highway. For each segment, number of lanes and traffic load based on average annual daily traffic (AADT) are also provided. Estimate the overall performance for a) ignoring traffic load, b) considering traffic load (Table 2.18).

**Table 2.19** Exercise 2.5 Database

| | | Condition (1–10) | Weight (%) | Weight (%) |
|---|---|---|---|---|
| Architectural | Floors | 8 | 25 | 20 |
| | Ceilings | 10 | 10 | |
| | Roof | 7 | 20 | |
| | Doors and Windows | 6 | 15 | |
| | Walls | 5 | 30 | |
| Structural | Foundation | 5 | 25 | 30 |
| | Columns | 6 | 20 | |
| | Beams | 6 | 20 | |
| | Stairs | 7 | 15 | |
| | Slabs | 4 | 20 | |
| Mechanical | HVAC | 3 | 20 | 20 |
| | Plumbing | 4 | 20 | |
| | Elevators | 5 | 20 | |
| | Escalators | 8 | 20 | |
| | Pumps | 4 | 10 | |
| | Tanks | 9 | 10 | |
| Electrical | Lighting | 7 | 20 | 20 |
| | Cabling | 6 | 20 | |
| | Panels | 7 | 20 | |
| | Switches | 8 | 20 | |
| | Emergency Power | 9 | 20 | |
| Safety | Fire Fighting System | 8 | 100 | 10 |

**Exercise 2.5**
A university building has recently been inspected by a consultant and the below table is the results of performance assessment and weighting for both component and sub-components levels. Estimate the overall performance of the building using both SAW and MAUT methods (Table 2.19).

## References

Amador-Jiménez, L., & Matout, N. (2014). A low cost solution to assess road's roughness surface condition for Pavement Management, TRB 93rd Annual Meeting, 2014-1-12 to 2014-1-16, Washington, DC United States (No. 14-3086).

American Public Transportation Association (APTA). (2013). Capital asset inventory and condition assessment. APTA SGR-TAM-RP00x-13. NW, Washington, DC, 20006-1215. https://www.apta.com/ (Nov 1, 2021).

American Society of Civil Engineers (ASCE). (2000). Guideline for condition assessment of the building envelope.

American Society of Civil Engineers (ASCE). (2017). 2017 report card for America's infrastructure, Retrieved on Sep.1, 2016 from http://www.infrastructurereportcard.org, (1 October 2018).

ASTM D6433-20. (2020). *Standard practice for roads and parking lots pavement condition index surveys*. ASTM International, www.astm.org
ASTM E1364-95. (2017). *Standard test method for measuring road roughness by static level method*. ASTM International, www.astm.org
ASTM E1926-08. (2015). *Standard practice for computing international roughness index of roads from longitudinal profile measurements*. ASTM International. www.astm.org
ASTM E2018-15. (2015). *Standard guide for property condition assessments: Baseline property condition assessment process*. ASTM International, www.astm.org
ASTM WBFS. (2000). *ASTM standards on whole building functionality and serviceability (WBFS)*. ASTM.
Balzer, G., & Schorn, C. (2015). *Asset management for infrastructure systems: Energy and water*. Springer.
Birolini, A. (2013). *Reliability engineering: Theory and practice*. Springer Science & Business Media.
Cafiso, S., Di Graziano, A., Goulias, D. G., & D'Agostino, C. (2019). Distress and profile data analysis for condition assessment in pavement management systems. *International Journal of Pavement Research and Technology, 12*(5), 527–536.
Canadian infrastructure report card (CIRC). (2016). Canadian infrastructure report card, Retrieved on Jan. 10, 2019 from http://www.canadainfrastructure.ca., (1 October 2018).
Chase, S.B., Adu-Gyamfi, Y., Aktan, A.E., and Minaie, E (2016). Synthesis of national and international methodologies used for bridge health indices. Federal Highway Administration https://www.fhwa.dot.gov/publications/research/infrastructure/structures/bridge/15081/15081.pdf
Chen, D., Hildreth, J., & Mastin, N. (2019). Determination of IRI limits and thresholds for flexible pavements. *Journal of Transportation Engineering, Part B: Pavements, 145*(2), 04019013.
Ciszewski, M., Giergiel, M., Buratowski, T., & Małka, P. (2020). Prototype of the pipe inspection robot. In *Modeling and control of a tracked mobile robot for pipeline inspection. Mechanisms and machine science, vol 82*. Springer. https://doi-org.lib-ezproxy.concordia.ca/10.1007/978-3-030-42715-3_7
Dhillon, B. S. (2004). *Reliability, quality, and safety for engineers*. CRC Press.
ERDC's Construction Engineering Research Laboratory (CERL) (ERDC- CERL). (2012). BUILDER™ sustainment management system. https://www.erdc.usace.army.mil/Media/Fact-Sheets/Fact-Sheet-Article-View/Article/476728/builder-sustainment-management-system/, (1 October 2018).
Federal Highway Administration (FWHA). (1995). *Recording and coding guide for the structure inventory and appraisal of the Nation's bridges*. U.S. Dept. of Transportation.
Fiix. (2020). Fiix software. https://www.fiixsoftware.com/ (Nov 1, 2021).
FPrmeC Solutions. (2019). Structural health monitoring for bridge structures. https://www.fprimec.com/structural-health-monitoring-for-bridge-structures/
Galasiu, A., Thompson, A., & Bergevin, P. (2019). *Literature review: Assessing building functional suitability—methods and tools*. National Research Council (NRC).
Ghodoosi, F., Bagchi, A., & Zayed, T. (2016). Reliability-based condition assessment of an externally restrained bridge deck system considering uncertainties in key design parameters. *Journal of Performance of Constructed Facilities, 30*(1), 04014189.
Hwang, C. L., & Yoon, K. (1981). *Multiple attribute decision making: Methods and applications: A state-of-the-art survey*. Springer-Verlag.
ISO. (2014). *55000 2014. Asset management. Overview, principles and terminology*. International Organization for Standardization.
ISO 11863. (2011). Buildings and building-related facilities - Functional and user requirements and performance - Tools for assessment and comparison,
ISO 15686. (2010). Buildings and constructed assets — service life planning, Part 10, When to assess functional performance.

ISO 2631-4. (2001). *Mechanical vibration and shock — Evaluation of human exposure to whole-body vibration — Part 4: Guidelines for the evaluation of the effects of vibration and rotational motion on passenger and crew comfort in fixed-guideway transport systems*. The Organization.

Jones, M., Williams, W., & Stillman, J. (2014). The evolution of asset management in the water industry. *Journal-American Water Works Association, 106*(8), 140–148.

Kaddoura, K. (2015). Automated sewer inspection analysis and condition assessment, (M.A. Sc. Thesis), Concordia University. https://spectrum.library.concordia.ca/980826/1/Kaddoura_MASc_S2016.pdf

Kaddoura, K., Zayed, T., & Hawari, A. H. (2018). Multiattribute utility theory deployment in sewer defects assessment. *Journal of Computing in Civil Engineering, 32*(2), 04017074.

Karam, M. (2016). Putting the customer at the heart of asset management, an interview with Kevin Young, Managing Director of Australia's Sydney Water Infrastructure Morality, insight magazine, Issue No. 8, KPMG, https://home.kpmg/xx/en/home/insights/2016/09/insight-magazine-past-editions.html. (1 October 2018).

Mohammadi, A., Amador-Jimenez, L., & Nasiri, F. (2018). Review of asset management for metro systems: Challenges and opportunities. *Transport Reviews*, 1–18.

Mohammadi, A., Amador-Jimenez, L., & Nasiri, F. (2020). A multi-criteria assessment of the passengers' level of comfort in urban railway rolling stock. *Sustainable Cities and Society, 53*, 101892.

Moradi, S., Zayed, T., Nasiri, F., & Golkhoo, F. (2020). Automated anomaly detection and localization in sewer inspection videos using proportional data modeling and deep learning–based text recognition. *Journal of Infrastructure Systems, 26*(3), 4020018.

Ontario Ministry of Transportation. (2013). Pavement design and rehabilitation manual, second edition, www.bv.transports.gouv.qc.ca/mono/1165561.pdf. (1 October 2018).

OtalPave. (2020). Smartphone app, TotalPave, https://totalpave.com/. (1 October 2018).

Phares, B. M., Washer, G. A., Rolander, D. D., Graybeal, B. A., & Moore, M. (2004). Routine highway bridge inspection condition documentation accuracy and reliability. *Journal of Bridge Engineering, 9*(4), 403–413.

Ryan, T. W., Mann, E., Chill, M., & Ott, T. (2012). *Bridge inspector's reference manual: BIRM*. U.S. Dept. of Transportation, Federal Highway Administration. FHWA NHI 12-050, https://www.fhwa.dot.gov/bridge/nbis.cfm

Saaty, T. L. (1980). *The analytic hierarchy process*. McGraw Hill.

Saaty, T. L. (2001). *Decision making with dependence and feedback: The analytic network process: The organization and prioritization of complexity*. RWS Publications.

Sayers, M. W., Gillespie, T. D., & Paterson, W. D. (1986). *Guidelines for the conduct and calibration of road roughness measurements, World Bank Technical Paper No. 46*. The World Bank.

Shahsavari, V., Mehrkash, M., & Santini-Bell, E. (2020). Damage detection and decreased load-carrying capacity assessment of a vertical-lift steel truss bridge. *Journal of Performance of Constructed Facilities, 34*(2), 04019123.

Su, Y. S., & Twu, S. H. (2020). A Real Time Fall Detection System Using Tri-Axial Accelerometer and Clinometer Based on Smart Phones. In K. P. Lin, R. Magjarevic, & P. de Carvalho (Eds.), *Future Trends in Biomedical and Health Informatics and Cybersecurity in Medical Devices. ICBHI 2019. IFMBE Proceedings, vol 74*. Springer. https://doi.org/10.1007/978-3-030-30636-6_19

SUEZ . (2021). SUEZ in North America. https://www.suez-na.com/. (Nov 1, 2021).

Tan S. (2015). Getting to know IRI and PCI: Their applications for local agencies, 2015 NWPMA Conference, October 20-23. http://nwpma-online.com/presentations/2015/iri-pci-tan.pdf. (1 October 2018).

Variableinc (2017). Retrieved on Dec 20, 2017 from http://www.variableinc.com/node1. (1 October 2018).

Zanini, M. A., Faleschini, F., & Casas, J. R. (2019). State-of-research on performance indicators for bridge quality control and management. *Frontiers in Built Environment, 5*, 22.

# Chapter 3
# Forecasting Infrastructure Performance

## 3.1 Introduction

As was explained in Chap. 1, infrastructure networks and facilities are created with the expectation of promoting human and economic development in a sustainable manner. For instance, the efficiency of any transportation network (and facility) can be related to its ability to support mobility (travel time), accessibility (to leisure, health, education, and work opportunities), and the safety of users while neutralizing or minimizing its environmental impacts.

The social and economic development promoted by the civil infrastructure can be linked to tangible key condition indicators, which properly can serve to support decision-makers to opt for the most sustainable (and cost-effective) strategies and actions to steer public networks of infrastructure and facilities in the desired direction.

For example, the very general goals of easy commuting to work and fast shipment of commodities and goods can be relevant to the most tangible KPIs of assets performance, travel time to work, and demand-to-capacity ratios. These KPIs can then be tracked through time as the years go by and (regression or structural equation) models can be developed to capture the historical behavior and forecast the future performance/condition of the system. This prediction would be the main element of long-term planning for infrastructure. Evidently, to forecast the future performance or condition of a system, one needs to determine (often through informed guess or simulation) the expected levels of relevant elements and characteristics underpinning the model behavior.

All these KPIs should be tracked to achieve sustainable and efficient solutions for infrastructure, particularly for long-term planning where maintaining, upgrading, and expanding will be considered. From an IAM perspective, there is less flexibility to address some KPIs such as travel time or demand-to-capacity ratio in decision-making for existing infrastructure, and they are more targeted in the expansion projects. That is possibly the main reason that these types of KPIs have been less



A. Mohammadi, L. Amador Jimenez, *Asset Management Decision-Making For Infrastructure Systems*, https://doi.org/10.1007/978-3-030-97614-9_3

concentrated in IAM efforts although recent works highlighted a potential to capture travel time and demand prediction for transit systems (Mohammadi et al., 2020) or demand capacity in storm pipelines (Amador et al., 2020) in IAM process.

For most networks of infrastructure, there are at least four main key performance areas of which the modeler wants to create a forecast of individual indicators: (1) The physical condition; (2) the level of safety offered to its users; (3) the demand-to-capacity ratio; and (4) the level of emissions/energy consumed during the operation of the system.

Among these four, the first one is the most common and classical approach used by decision-makers to build IAM platforms, and other areas are still under study by scholars toward performance-based planning. Thus, this book concentrates on forecasting physical conditions; however, discussions are also included to elaborate on other potential solutions. The performance and physical condition of any infrastructure or facility are always at the center of IAM procedures. Not only the current level of the condition can be measured (Chap. 2) but the impact from environmental cycles and usage demand can be considered to predict future levels of accumulated damage and degradation of each of the components up to the point in which they are not capable of providing the intended service either because failure is imminent or adequate operation is not possible anymore. Think for example of a pavement full of potholes; although a vehicle can maneuver through it, this cannot be considered adequate operation. Assets naturally degrade by time and under operation and this must be captured in the decision-making process by forecasting future conditions.

Predicting how the levels of condition and any indicator relevant to a specific asset change across time is fundamental to making appropriate and timely decisions. Forecasting opens the doors to improved planning because the impact of decisions at any point in time can be measured and assessed through the KPIs and back into the general goals to be accomplished. Enabling a recursive and dynamic cost-benefit analysis that ensures minimum levels of condition for any KPI are met through time.

## 3.2 Homogeneous Groups and Database Query

Most infrastructure data are held at local databases and queries are necessary to extract sets of data from assets or segments with very similar characteristics (called "homogeneous groups") and expected to have a close condition prediction trend. There are also approaches in the literature where self-organizing-maps (SOMs) or K-means (Mathavan, 2014) have been proposed to identify such homogeneous groups (HGs). Asset managers are expected to identify all factors contributing to asset condition prediction (i.e., deterioration) classifying them in HGs. Then the deterioration models will be developed separately for each group. Increasing the number of HGs may improve the accuracy of the deterioration model; however, it would be challenging to provide enough data to develop a model for each group and one needs to ensure there is a notable difference between groups.

### 3.2.1 *Sample 1, Homogeneous Groups (HGs) for a Road Pavement*

For a road pavement, ideal HGs will encompass all pavements within the same climatic region, observing similar levels of annual accumulated truck traffic loads, with similar structural capacity and made out of the same materials.

Table 3.1 presents a sample of the main factors contributing to the deterioration of the pavement. ESAL stands for equivalent single axle loads as a common traffic load indicator and SN is structural number, which represents structural strength for roads.

Based on this classification, some examples of HGs are provided as below:

- Group 1: Arterial, High Traffic Loading, Flexible, 4 < SN < 5, Dry and Non-freeze
- Group 2: Highway, High Traffic Loading, Rigid, SN > 5, Dry and Non-freeze
- Group 3: Collector, Low Traffic Loading, Flexible, SN < 4, Dry and Non-freeze
- Group 4: Collector, Low Traffic Loading, Flexible, SN < 4, Wet, and Freeze.

Mature PMS uses more detailed HGs. Table 3.2 presents another example for pavements. As the reader notices, the ESALs per year variation is responsible for groups "Flexible Hot Mixed Asphalt (HMA) 1" through "Flexible HMA 3"; the structural capacity measured by the "SN" is responsible for the variation between "Flexible HMA 1" and "Flexible HMA 4"; the climatic region is responsible for the variation between groups "Flexible HMA 1" and Flexible HMA 7″, the material type is responsible for the variation between groups "Flexible HMA 1" through "Flexible HMA 9" and "Rigid PC 1" and subsequent groups.

Advanced tools and clustering techniques such as the K-means technique or self-organizing map (SOM) can be used to identify the optimal number of groups, and the range of each characteristic (ESALs per year, SN, climatic region, material type). Both techniques create a cloud of points for each characteristic (principal component) measuring how important it is for a given condition indicator. The K-means imposes a hard clustering approach where each observation must belong to a given

**Table 3.1** One Sample Approach for Defining HGs for Pavement

| Main Deterioration Characteristics for Pavement | | | | |
|---|---|---|---|---|
| Function | Traffic Loading | Structure 1 | Structure 2 | Environmental Condition |
| Highway | High (ESAL>100,000) | Rigid | SN > 5 | Dry and non-freeze |
| Arterial | Medium (40,000 < ESAL <100,000) | Flexible | 4 < SN < 5 | Dry and non-freeze |
| Collector | Low (ESAL <40,000) | Flexible | 3 < SN < 4 | Dry and non-freeze; wet and freeze |
| Local | Very Low (ESAL<10,000) | Flexible | 2 < SN < 3 | All |

**Table 3.2** Another Example for Pavement HGs

| Group Number | ESALs per year | Structural Capacity | Climatic Region. Thornthwaites Moisture Index |
|---|---|---|---|
| Flexible HMA 1 | 100,000 to 800,000 | 3.0 < SN < 4.0 | M.I = 80%, Cold |
| Flexible HMA 2 | 800,001 to 1,600,000 | 3.0 < SN < 4.0 | M.I = 80%, Cold |
| Flexible HMA 3 | 1,600,000 to 3,200,000 | 3.0 < SN < 4.0 | M.I = 80%, Cold |
| Flexible HMA 4 | 100,000 to 800,000 | 4.0 < SN < 5.0 | M.I = 80%, Cold |
| Flexible HMA 5 | 800,001 to 1,600,000 | 4.0 < SN < 5.0 | M.I = 80%, Cold |
| Flexible HMA 6 | 1,600,000 to 3,200,000 | 4.0 < SN < 5.0 | M.I = 80%, Cold |
| Flexible HMA 7 | 100,000 to 800,000 | 3.0 < SN < 4.0 | M.I = 60%, Cold |
| Flexible HMA 8 | 800,001 to 1,600,000 | 3.0 < SN < 4.0 | M.I = 60%, Cold |
| Flexible HMA 9 | 1,600,000 to 3,200,000 | 3.0 < SN < 4.0 | M.I = 60%, Cold |
| Rigid PC 1 | 1,100,000 to 2,200,000 | 4.5 < SN < 5.0 | M.I = 80%, Cold |
| ... | ... | ... | ... |

**Table 3.3** Sample Dataset for Clustering

| ESALs | SN | ChangeIRI[a] |
|---|---|---|
| 100,000 | 3 | −0.25 |
| 110,000 | 3.1 | −0.27 |
| 250,000 | 2.9 | −0.39 |
| 240,000 | 5 | −0.01 |
| 330,000 | 5.1 | −0.015 |
| 350,000 | 4.9 | −0.01 |
| 400,000 | 3.8 | −0.07 |
| 500,000 | 3.3 | −0.1 |
| 600,000 | 3.8 | −0.08 |
| 650,000 | 3.4 | −0.12 |

[a]Average Annual drop in IRI m/km

group, meanwhile the SOM uses a soft clustering approach where observations can belong to two clusters, and this is particularly important for observations located at the boundary of two clusters. Both techniques are available in common statistical software packages such as MATLAB (MathWorks, 2020), RStudio (RStudio, 2020), IBM-SPSS (IBM, 2020), Minitab (Minitab, 2020).

Let us assume you are presented with the following data (Table 3.3).

In Minitab, go to STATS < Multivariable < Cluster K-Means, select all three variables (ESALS, SN, ChangeIRI), use 3 clusters for this example. The software

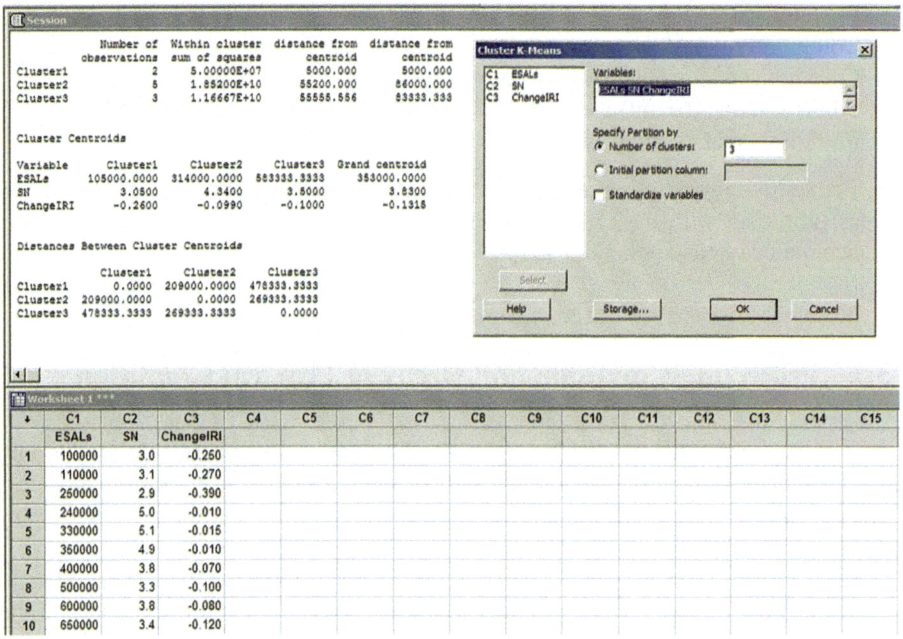


**Fig. 3.1** Clustering HGs with Minitab

output is presented in Fig. 3.1. The algorithm identified three clusters at average ESALs of 105,000; 314,000; and 583,333 with corresponding structural numbers of 3.05; 4.34; and 3.5 and annual drop in IRI of −0.26 m/km; −0.099 m/km and − 0.100 m/km per year. From these cluster centers, one can build ranges by splitting the variable ranges into three subsets.

Identified deterioration characteristics and HGs are also key factors in defying road segments for the purpose of AM. From an operational and practical perspective, road agencies should split a road into shorter pieces to implement maintenance interventions; however, this must be done to ensure pavement with the same or similar deterioration characteristics are planned together. For instance, if a road between intersections 1 to 4 has two different traffic loads or structural types, then this part (i.e., intersections 1 to 4) cannot be considered as one segment and should be divided accordingly.

### 3.2.2 *Sample 2, HGs for Pipeline Networks*

A similar approach can be used for other infrastructure to identify HGs. The possible main factors contributing to the deterioration of pipelines can be:

- Pipe material
- Diameter range

- Pressure range (internal and external)
- Commodity type (corrosion-wise)
- Location (underground or aboveground)
- Soil (for underground)
- Environmental conditions

The following sections explain in detail each of the main key condition prediction areas (initiating with the classical one) and will further discuss individual performance indicators as applicable to specific infrastructure assets.

## 3.3 Projecting Condition and Decay of Physical Condition

As was disused earlier, degradation or deterioration is a process that affects any asset: for metallic materials like bridges, this means corrosion; for concrete elements like decks of bridges, delamination; and for asphalt pavements, rutting, cracking, or raveling.

Each performance indicator characterizing the condition of an asset must relate to a specific type of damage because this facilitates selecting a countermeasure in the form of treatment, action, or intervention that either mitigates (reduce but not fully recover) or eliminates the deficiency observed bringing back the asset to a brand-new like state.

Fig. 3.2 presents a typical deterioration trend where asset loses condition (0–100 scale) through the lifespan. As can be seen, it follows a smooth decay at first while the asset is still new; however, following that, the trend of losing condition would be much faster where asset condition drops suddenly (i.e., aging). Finally, when the asset passes condition thresholds (i.e., the minimum acceptable condition) and reaches the end of service life, the decay trends will be almost constant. Identifying

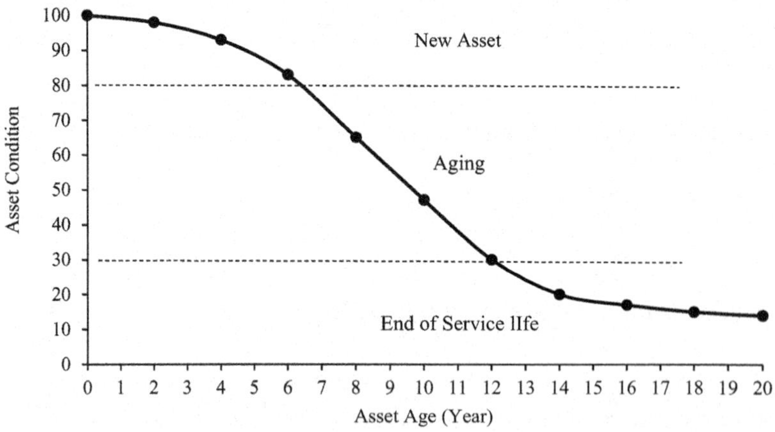


**Fig. 3.2** Typical trends of deterioration for asset

these milestones is key for intervention timing. For instance, applicable maintenance actions for new assets (e.g., preventive maintenance) are often cheaper and more cost-effective while a more expensive major and corrective intervention might be the only solution for assets in the aging zone. Passing condition thresholds, the cost of replacement or renovation would be dramatically higher. It is critical for decision-makers to be aware when an asset enters the aging period or is close to the end of service life.

Therefore, the role of a deterioration model is more than just projecting condition through life span while identifying the decay trend in the whole service life. It should be mentioned that the decay model for some infrastructure may follow other trends such as exponential or even linear depending on the nature of the asset as well as data availability. For mechanical or electrical assets, the end of service life could be failure time, and reaching performance equals zero.

Condition prediction is about estimating how values for each condition indicator will change over time. The development of models to project condition relies heavily on the understanding of the degradation mechanisms and contributing factors impacting in an isolated and combined manner and sourced out of natural or manmade loading mechanisms ranging. These factors cover from user's demand (truck traffic spectrum, hydraulic demand for water systems) to climatic exposure (environmental freeze-thaw cycles, moisture levels, wind, earthquakes), and the quality of the structure and materials utilized (the structural capacity and constructive methods employed).

The prediction of asset conditions requires mathematical models, which vary in degree of complexity, depending on the availability of information and the skills of the modeling team. Models could launch as simple as linear or non-linear regression with best-fit estimations for historically observed levels of a given indicator, for example, cracking on a pavement surface. Regression is easy to understand and implement; however, the problem with a best-fit regression is that the obtained equation depicts only what was observed in the past. If any of the previous contributing factors change (per instance due to climate change the number of precipitations increases, or because of urbanization the flow of water increases), then the best fit model is not applicable anymore and the team needs to redo the analysis.

There are multiple ways to classify deterioration models and the methodology behind them. The decay model can be deterministic or probabilistic excluding or including uncertainty. Models can be empirical or mechanistic or a combination of both. Finally, there are several AI methods recently used to develop deterioration models. Fig. 3.3 presents a general view of the most common approaches to classify deterioration models.

One of the most famous equations of road pavement deterioration can be found in the highway design manuals and is based upon experiments conducted in the 1980s (Paterson & Attoh-Okine, 1992). The equation includes the IRI of a brand new road ($IRI_0$), which can vary depending on the road function and the construction quality; an environmental factor that characterizes the moisture index ($m$) as recommended by Thornthwaite's ($m = 0.07$ for moisture index of 80, $m = 0.074$ for moisture index

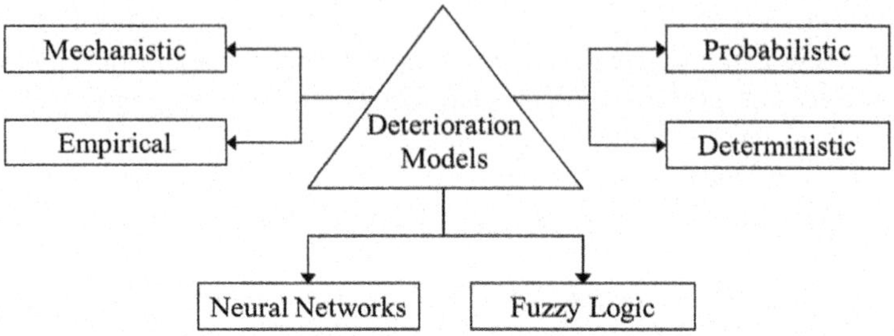


**Fig. 3.3** The most common classification for deterioration model

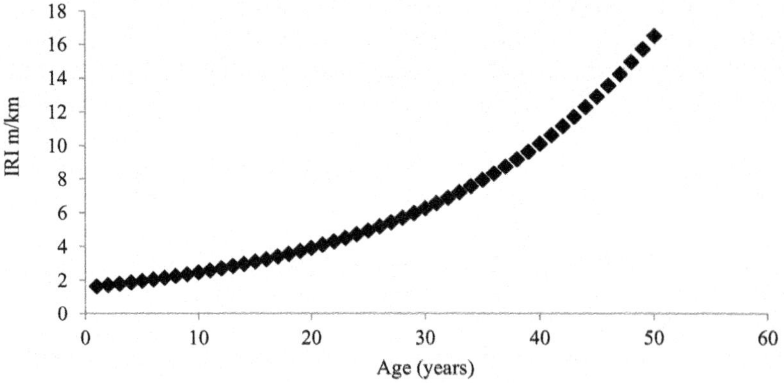


**Fig. 3.4** Pavement Surface Condition—International Roughness Index (m/km)

of 100, etc.) (Zareie et al., 2016); the accumulated number of truckloads in millions (ESAL); and the pavement structural capacity (SNC). As seen in Eq. 3.1 and Fig. 3.4, some coefficients came from a calibration of this equation using real observations.

$$\mathrm{IRI}_t = e^{mt}\left[\mathrm{IRI}_0 + \alpha(\mathrm{SNC})^{-5} \times \mathrm{ESAL}_t\right] \tag{3.1}$$

These equations for deterioration are known as empirical models because they don't really reach back to the mechanics of materials (elastic or plastic behavior) estimating the materials' stresses and strains, and from there predicting the condition. Various prediction models follow this paradigm and simply use reasonable assumptions to forecast the levels of causal factors at multiple points in time and input them into the equation of deterioration.

Other models called mechanistic track down the material responses (stresses and strains) and relate them directly to the amount of permanent deformation suffered by the asset. In some cases, the accumulation of damage is still done through calibrated expressions even though the estimation of the response is mechanistic. For this reason, such models are known as mechanistic-empirical.

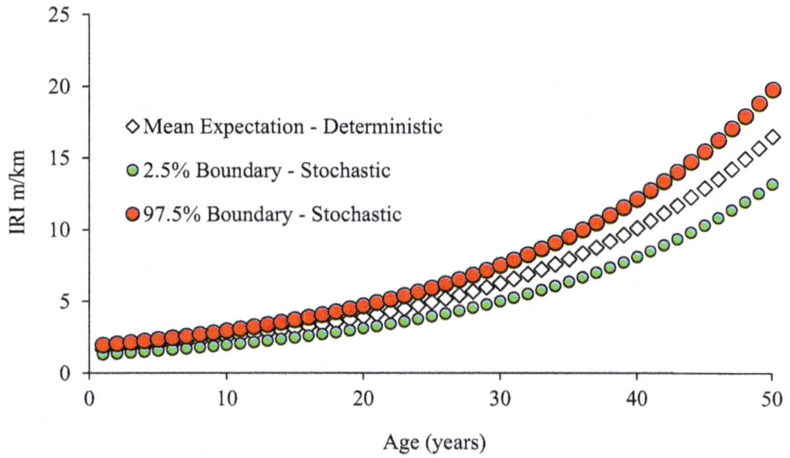


**Fig. 3.5** Deterministic versus Stochastic

Nevertheless, the method that follows a simple forecast of mean expectation of the KPI value for any point in time is called deterministic, which means the user only sees a line (or curve) plotted for the given indicator through time. Probabilistic (i.e., Stochastic) models bring one additional layer of sophistication and, in addition to the mean, they also include enveloping confidence intervals around the mean expectation line. The confidence intervals can vary depending on the degree of reliability that the user desires, being the most common 95% (Fig. 3.5), but in some cases being perfectly normal to accept a 75% or less (for instance pavement design of low-volume roads can use as little as 60% confidence intervals).

Stochastic models can also produce a probability plot for the range of variation around a given causal factor or explanatory variable associated with the degradation of an asset (Fig. 3.6). The probability plot could, in most cases, be assumed to follow a near-normal distribution with a bell-shaped curve and having its mean halfway through, adding cutting lines for the tails, which reflect again the confidence interval, but this time for the likely value of the calibration value to be used on the empirical components discussed earlier.

AI methods such as Artificial Neural Network (ANN) have been also tested to model complex and nonlinear relationships between inputs; however, the need for a large amount of data and black box in this method limits the transparency of the solution path (Ens, 2012).

Asset managers need to select suitable methods considering several factors including data availability, types of assets, and AM maturity. Data availability is often the most challenging criterion. Therefore, the coming Sects. 3.3.1, 3.3.2, and 3.3.3 are designed to represent three practical situations respecting data availability, and then potential methods in place are discussed. Asset engineers may have to start without historical data then by collecting records over time move to more accurate solutions.

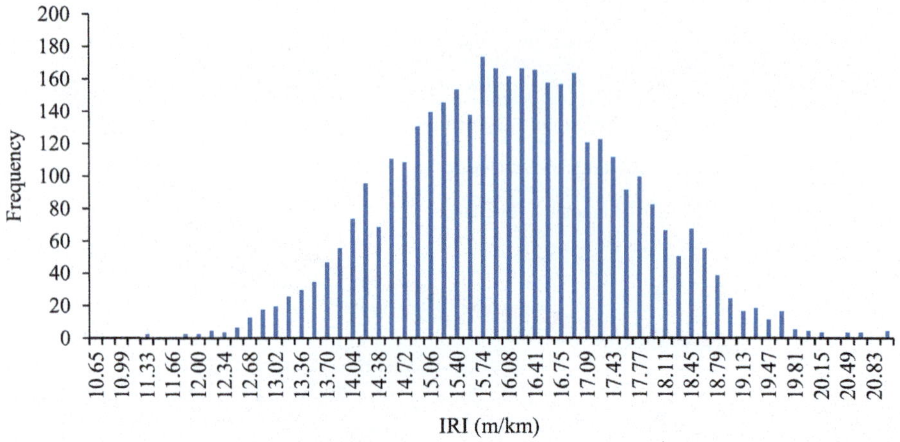


**Fig. 3.6** Stochastic Probability on year 50

## 3.3.1 *No Time-Series Data*

Being proactive through long-term plans is essential to running a mature business. Therefore, in the event that no historical time-series data exists, deterioration models can be developed using approximate models. However, decision-makers must be cautious and pick smart approaches while the decay model should be actively reviewed incorporating newly collected data. These are potential solutions:

- Using empirical and mechanistic models
- Borrowing a deterioration model from other locations
- Useful life recommendations
- Weibull distribution

### 3.3.1.1 Using Empirical and Mechanistic Models

Empirical models use directly measurable characteristics of the asset and the loads (demand) experienced to estimate the degree of damage accumulation across time. Various empirical models can be found in the literature while the most popular are those for Pavements Serviceability (Hall & Correa-Muñoz, 1999).

As was discussed earlier, World Bank international roughness equation (Eq. 3.1) for pavement is an example of well-known global models. The number of ESALs will be accumulated using the expected traffic growth and vehicle classification. HGs will be formed, and one deterioration curve created for each group. For example, HGs for strong pavement structure (SN > 6), fair structure (4 < SN < 6), and weak structure (SN < 4) can be created. Traffic intensity will be used as other criteria to further break down the pavement HGs into high traffic intensity, medium traffic intensity, and low traffic intensity. Climate regions would also be used if applicable.

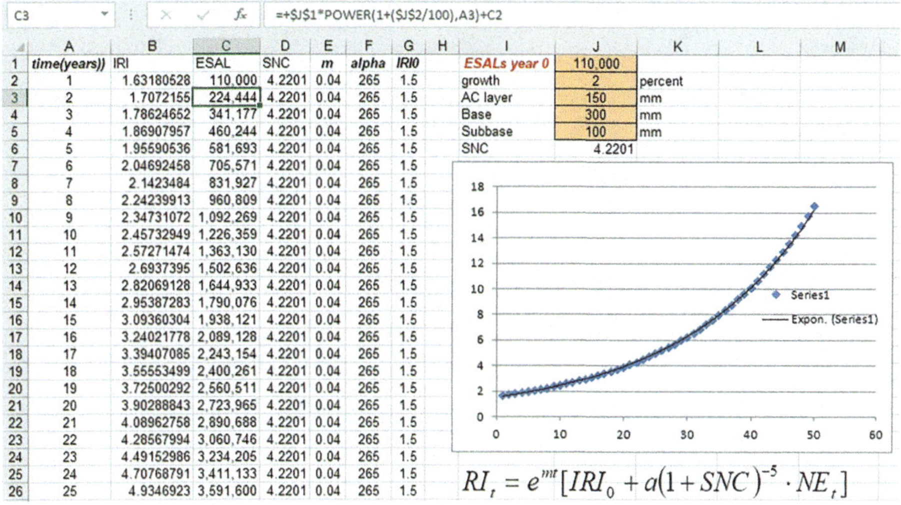
C3 =+$J$1*POWER(1+($J$2/100),A3)+C2

| | A | B | C | D | E | F | G | H | I | J | K |
|---|---|---|---|---|---|---|---|---|---|---|---|
| 1 | *time(years))* | IRI | ESAL | SNC | *m* | *alpha* | *IRI0* | | *ESALs year 0* | 110,000 | |
| 2 | 1 | 1.63180528 | 110,000 | 4.2201 | 0.04 | 265 | 1.5 | | growth | 2 | percent |
| 3 | 2 | 1.7072155 | 224,444 | 4.2201 | 0.04 | 265 | 1.5 | | AC layer | 150 | mm |
| 4 | 3 | 1.78624652 | 341,177 | 4.2201 | 0.04 | 265 | 1.5 | | Base | 300 | mm |
| 5 | 4 | 1.86907957 | 460,244 | 4.2201 | 0.04 | 265 | 1.5 | | Subbase | 100 | mm |
| 6 | 5 | 1.95590536 | 581,693 | 4.2201 | 0.04 | 265 | 1.5 | | SNC | 4.2201 | |
| 7 | 6 | 2.04692458 | 705,571 | 4.2201 | 0.04 | 265 | 1.5 | | | | |
| 8 | 7 | 2.1423484 | 831,927 | 4.2201 | 0.04 | 265 | 1.5 | | | | |
| 9 | 8 | 2.24239913 | 960,809 | 4.2201 | 0.04 | 265 | 1.5 | | | | |
| 10 | 9 | 2.34731072 | 1,092,269 | 4.2201 | 0.04 | 265 | 1.5 | | | | |
| 11 | 10 | 2.45732949 | 1,226,359 | 4.2201 | 0.04 | 265 | 1.5 | | | | |
| 12 | 11 | 2.57271474 | 1,363,130 | 4.2201 | 0.04 | 265 | 1.5 | | | | |
| 13 | 12 | 2.6937395 | 1,502,636 | 4.2201 | 0.04 | 265 | 1.5 | | | | |
| 14 | 13 | 2.82069128 | 1,644,933 | 4.2201 | 0.04 | 265 | 1.5 | | | | |
| 15 | 14 | 2.95387283 | 1,790,076 | 4.2201 | 0.04 | 265 | 1.5 | | | | |
| 16 | 15 | 3.09360304 | 1,938,121 | 4.2201 | 0.04 | 265 | 1.5 | | | | |
| 17 | 16 | 3.24021778 | 2,089,128 | 4.2201 | 0.04 | 265 | 1.5 | | | | |
| 18 | 17 | 3.39407085 | 2,243,154 | 4.2201 | 0.04 | 265 | 1.5 | | | | |
| 19 | 18 | 3.55553499 | 2,400,261 | 4.2201 | 0.04 | 265 | 1.5 | | | | |
| 20 | 19 | 3.72500292 | 2,560,511 | 4.2201 | 0.04 | 265 | 1.5 | | | | |
| 21 | 20 | 3.90288843 | 2,723,965 | 4.2201 | 0.04 | 265 | 1.5 | | | | |
| 22 | 21 | 4.08962758 | 2,890,688 | 4.2201 | 0.04 | 265 | 1.5 | | | | |
| 23 | 22 | 4.28567994 | 3,060,746 | 4.2201 | 0.04 | 265 | 1.5 | | | | |
| 24 | 23 | 4.49152986 | 3,234,205 | 4.2201 | 0.04 | 265 | 1.5 | | | | |
| 25 | 24 | 4.70768791 | 3,411,133 | 4.2201 | 0.04 | 265 | 1.5 | | | | |
| 26 | 25 | 4.9346923 | 3,591,600 | 4.2201 | 0.04 | 265 | 1.5 | | | | |



**Fig. 3.7** Simplified IRI Deterministic Model. (Patterson and Attoh-Okine 1992)

Consider an example where we have an SN of 4.2201, and we know there are 110,000 ESALs in the current year, and the growth of ESALs is expected at 2% per year while m is 0.04 and $\alpha$ is assumed to be equal to 265. Fig. 3.7 shows how the model can be plotted using an Excel spreadsheet.

Mechanistic-empirical models utilize the theory of elasticity to estimate critical strains on the material under a given load (mechanistic component) and accumulate damage through calibrated empirical expressions. This section presents one example of mechanistic-empirical models for the pavement riding comfort index (RCI) (Ontario Ministry of Transportation, 2013). More examples can be reviewed in Appendix 1.

RCI is based on the Ontario Model called OPAC 2000 where the pavement performance is measured while

$$\mathrm{RCI}_t = \mathrm{RCI}_0 - \Delta\mathrm{RCI}_t - \Delta\mathrm{RCI}_e \tag{3.2}$$

where $\mathrm{RCI_o}$ is the initial RCI and delta_RCI_*t* is the performance loss due to traffic, delta_RCI_*e* is the performance loss due to the environment.

The equivalent granular thickness ($H_e$) is calculated by transforming the pavement layers thickness using strength coefficients ($a1*H_1 + a_2*H_2 + a_3*H_3$), also known as granular base equivalency factors. This results in a two-layer pavement structure to which the deflection to any given load ($P$) can be estimated using the modulus of subgrade reaction in MPa ($M_s$) and the modulus of the equivalent granular base material ($M_2$) which tends to average near 345 Mpa, using the following equations:

**Table 3.4** Ontario *a* and *b* Coefficients

| Parameter | Southern Ontario | Northern Ontario |
|---|---|---|
| "*a*" | −0.0329 | −0.0415 |
| "*b*" | 12.7211 | 10.5478 |
| $R^2$ (Coef. of determination) | 0.707 | 0.866 |

$$W_s = \frac{1000 * P}{2 * M_s * Z * \sqrt{a + \left(\frac{a}{Z}\right)^2}} \tag{3.3}$$

where $a$ is the radius of loaded area (approximately 163 mm for the imprint of a dual tire load, $P$ is 40 kN for a dual tire carrying the standard axle load of 80kN, and $Z$ is defined as follows:

$$Z = 0.9 * H_e * \left(\frac{M_2}{M_s}\right)^{\frac{1}{3}} \tag{3.4}$$

The loss of Riding comfort index is defined from the deflection and the number of 80 K ESAL ($N$)

$$\Delta \mathrm{RCI}_t = 2.4455 * \left(3.7239 * 10^{-6} * W_s{}^6 * N\right) + 8.805 * \left(3.7239 * 10^{-6} * W_s{}^6 * N\right)^{\frac{1}{3}} \tag{3.5}$$

The calculation of the environmental loss is based on the initial RCI, the deflection ($W_s$) the pavement age ($Y$), and constants $a$ and $b$ as follow:

$$\Delta \mathrm{RCI}_e = \mathrm{RCI}_0 * \left[1 - \left(\frac{1}{1 + b * W_s}\right)\right] * \left(1 - e^{a*Y}\right) \tag{3.6}$$

For example, in Ontario, the coefficients $a$ and $b$ take the following values (Table 3.4):

#### 3.3.1.2 Borrowing a Deterioration Model

Some industry best practices also provide decay models for different assets based on their database, which can be used as benchmarks to fill this gap. Fig. 3.8 shows one example of decay models for 40-foot buses provided by the NCTR, 2016. A database of many bus agencies across the US has been used to develop models. These deterioration curves show how 40-foot buses lose condition over the life cycle and reach the end of service life (condition 2.5 in this guideline). Three models are proposed based on implemented preventive maintenance (PM) scenarios.

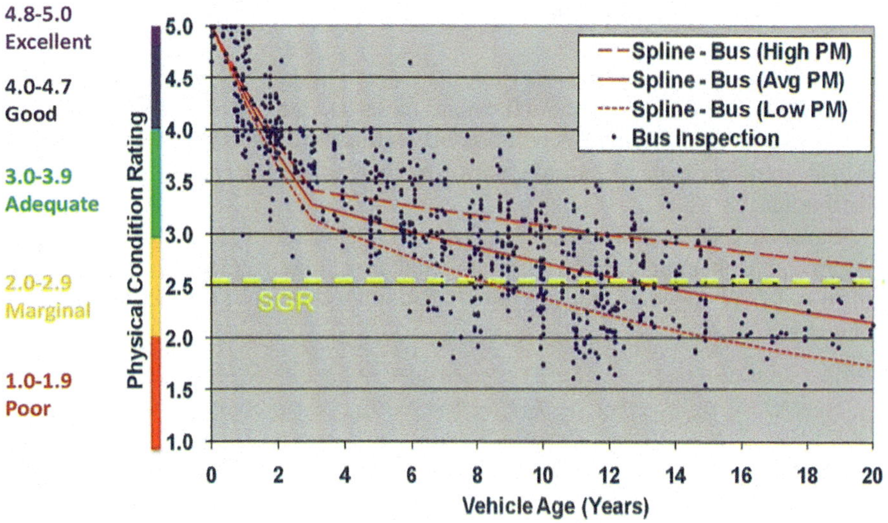


**Fig. 3.8** Sample decay model for bus fleet. (Adapted from NCTR, 2016)

Depending on the type of asset, this approach only is recommended for assets with similar characteristics (e.g., the similar type of vehicles and annual mileage per bus) while always there are hidden factors that cannot be fully investigated. For this reason, the borrowing deterioration model always brings the risk of over- or under-estimating asset service life and causing misleading.

#### 3.3.1.3 Useful Life Recommendations

Industry best practices also provide useful life recommendations for assets reflecting the age that assets typically reach the end of useful life (here are some samples: Infrastructure Canada, 2016; RTA, 2014; Toronto Hydro-Electric System, 2009; and FTA, 2008). For mechanical and electrical-based assets like a pump or transformer, normally, manufacturers provide this information (useful life per year or hours of operation). The manufacturer may also advise on preventive-corrective maintenance actions and timing. These are data-driven recommendations, which are helpful for asset owners, however, should be calibrated to address real-world practice for each site.

Another valuable source to recommend on asset useful life is subject matter expert. Experts judgment and lessons learned on the useful life of assets, timing for corrective actions, and associated costs are key information to build AM platform and to plan for the future. This can be used even to calibrate best practices and manufacture recommendations and to capture specific internal (e.g., agency practice) and external (e.g., environmental conditions) factors as well as hidden players.

### 3.3.1.4 Weibull Distribution

By identifying only useful life, decision-makers can find out when an asset should be replaced while will be blind about the trend of deterioration over asset lifecycle, and this gap limits applying advanced trade-off methods to find optimal solutions (will be discussed in Chap. 5). An alternative method to fill this gap is using Weibull distribution.

Semaan (2011) and Gkountis (2014) used this approach for a metro management system and Mohammadi et al., 2020 applied to asset management systems for urban railways to fill the data gaps. Weibull Cumulative Density Function (CDF) is defined as:

$$F(t) = 1 - e^{-\left(\frac{t-\gamma}{\delta}\right)^{\beta}} \tag{3.7}$$

Where:

$\beta, \gamma$, and $\delta$ are shape, location, and scale parameters, respectively. Then, reliability $R(t)$ (i.e., condition $C(t)$) function of time ($t$) (e.g., age) could be presented as:

$$R(t) = 1 - F(t) = C(t) = e^{-\left(\frac{t-\gamma}{\delta}\right)^{\beta}} \qquad 0 \leq C(t) \leq 1 \tag{3.8}$$

Using the shape parameter ($\beta$) equals three results in the most typical trend of asset deterioration presented in Fig. 3.2 (Semaan, 2011). Meanwhile, in the time zero, it could be assumed that $C$ (0) = 1, as the asset is brand new; therefore, the location parameter ($\gamma$) would be zero.

$$C(t) = e^{-\left(\frac{t}{\delta}\right)^{3}} \qquad \ln\left(C(t)\right) = -\left(\frac{t}{\delta}\right)^{3} \tag{3.9}$$

Now, we need a pair of age (i.e., apparent age) and the corresponding condition to solve the equation finding the scale parameter ($\delta$). Recommended service life (i.e., age as the end of asset life) and corresponding condition level can be also used to find ($\delta$) in case of having no observation.

**Example 3.1**

Let us assume condition 2 out of 5 (or $C(t)$ = 0.4) is the minimum acceptable condition level for lighting poles and those that are past this level should be replaced to avoid failure. Lesson learned from the past indicates that on average this asset will reach the end of service life at age 40. Using Weibull distribution, $\delta$ can be estimated (Eq. 3.10), and then the deterioration model can be developed as Eq. 3.11 for this pair of asset condition and age. Then asset conditions for the entire lifecycle can be forecasted (Fig. 3.9).

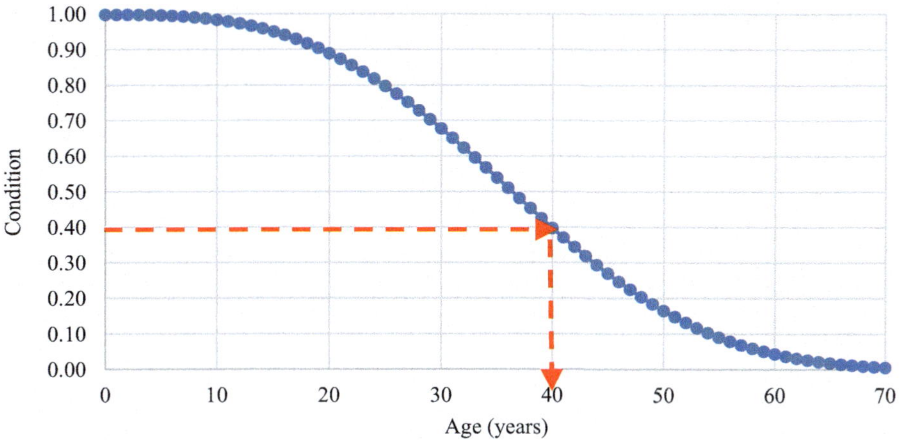


**Fig. 3.9** Weibull deterioration model for a pair of age and condition (Example 3.1)

$$0.4 = e^{-\left(\frac{40}{\delta}\right)^3} \qquad \delta = 41.18 \tag{3.10}$$

$$C(t) = e^{-\left(\frac{t}{41.18}\right)^3} \qquad t: 0, 1, 2, 3, \ldots\ldots\ldots \tag{3.11}$$

In a situation when the team never experienced replacing this kind of asset nor has a trustable benchmark, the solution can be using condition assessment results. If we know asset age (i.e., apparent age) and recent assessment provides a condition for the same asset, then the same method can be applied using this pair of age and condition to develop a deterioration model. By developing the decay trend, this time model provides an approximate useful life.

**Example 3.2**
Another asset is currently at age 20 and assessment shows condition 0.8. $\delta$ can be calculated (Eq. 3.10) and decay trend can be developed (Fig. 3.10). Using this model, the corresponding age for condition equals 0.4 (assumed as the end of service life) would be age 32, which means useful life per year for this asset.

**Example 3.3**
A customized and more accurate approach would be collecting more pair points using condition assessment (can be multiple observations) as well as recommended useful life and, finally, an average of the scale parameter can be used to develop a more realistic deterioration model. This helps to be risk-neutral rather than optimistic or pessimistic. Fig. 3.11 shows another example where there are two sample data as described below (Table 3.5). One can use an average model in this case.

Introduced methods in this section are intended to fill gaps while organizations are expected to enhance decay models by collecting data and incorporating historical records. The next sections discuss other approaches where this information is available.

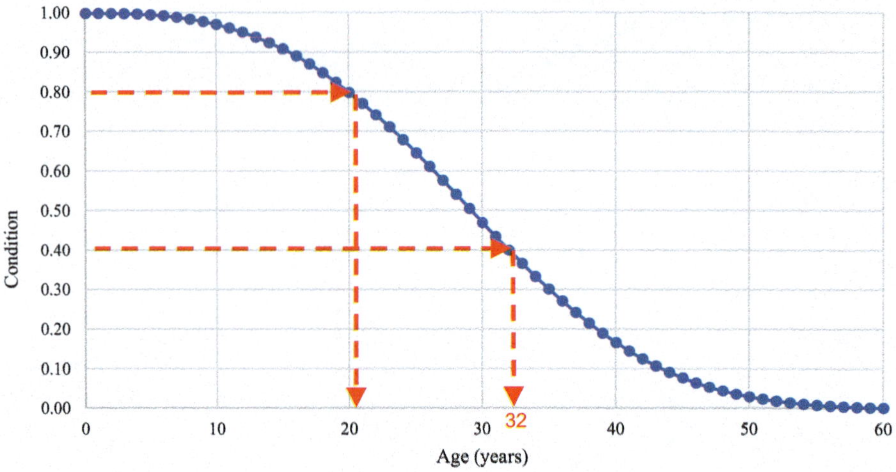


**Fig. 3.10** Weibull deterioration model for a pair of age and condition (Example 3.2)

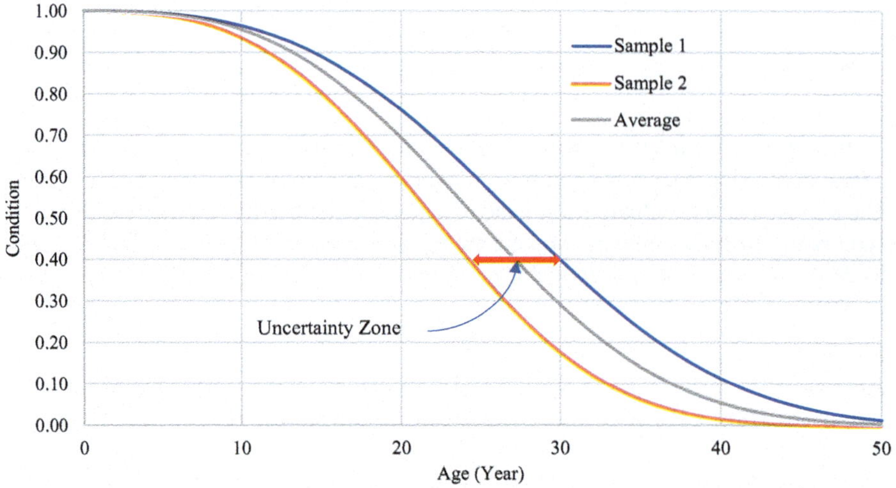


**Fig. 3.11** Weibull deterioration model for multiple pair of age and condition (Example 3.2)

**Table 3.5** Two Sample Pairs

| Sample | Age | Condition |
|---|---|---|
| 1 | 30 | 0.4 |
| 2 | 20 | 0.6 |

### 3.3.2 *Two-Point Time-Series Data*

#### 3.3.2.1 Deterministic Model

A model to handle situations with few time-series points was first proposed by Amador-Jimenez and Mrawira (2009) and further documented by various case studies including Costa Rica National Road Network (Amador-Jimenez & Mrawira, 2011a, 2011b), Ecuador's province of El Oro (Amador-Jimenez & Serrano, 2017), and Montreal (Mohammadi et al., 2019).

This method develops one curve for each homogeneous group. For instance, observations from various road segments within a given HG are considered as having a cross-section of individuals at various points of their lifespans. Therefore, a further grouping of all pavements is conducted to cluster them into 3 or 4 average individuals at initial, middle, and advanced conditions within their lifespan. For example, for a 3-point lifespan we refer to pavements in very good condition as young individuals, the adults will be pavements in fair condition, and elders will be pavements in poor condition.

For each of the 3 individuals, we estimate the average condition in the initial year when data is available and the average condition in the subsequent year. In this way, only 2 time-series data points are required. Given the short distance between time points, one can assume a linear slope for the change in condition and use it to find out the corresponding pairs of coordinates ($x_1$ = initial time, $y_1$ = initial average condition indicator; $x_2$ and $y_2$ the same for the subsequent year). The best-fitted prediction curve is used then with the 6 data points (or more) to draw a deterioration curve for the corresponding homogeneous group.

The deterioration model is developed based on two consequent observations and the time interval between observations could be one or more years, but shorter periods (one or two years) usually increase the model accuracy as they better capture the non-linear nature of pavement deterioration. A general methodology proposed by Mohammadi et al. (2019) is presented here and an easy, fast, and applicable approach for agencies is proposed to develop their own deterioration model. The formulation could be easily customized for one or more than one-year intervals to be matched with available data. Having the ability to adjust the interval is key to ensure local circumstances are considered while deciding the period of time needed for the assessment of the road network condition. The following steps should be taken to find the best-fitted curve:

Step1: Select a condition indicator (e.g. PCI or BCI), define a range for each state of the lifespan (i.e., good, fair, poor, etc.), and categorize pavement segments in ranges according to their base year (first) level of condition. For instance, PCI range of good (100-70), fair (70-50), poor (50-30), very poor (30-0).

Step 2: Estimate the average condition for each range in the initial year and subsequent time point.

Step 3: Bring the average value of each condition range into the calculator shown in Table 3.6. The calculator shows elements for a model with four subgroups, and the

**Table 3.6** Calculator for Developing the Best-Fitted Curve

| Defined Groups Range | Variable Age | Observations | Average PCI | Estimated Age | Age Difference |
|---|---|---|---|---|---|
| $PR_1$ | $V_1$ | | | | |
| | | First | $AP_{11}$ | $A_{11}$ | $A_{21}–A_{11}$ |
| | | Second | $AP_{21}$ | $A_{21}$ | |
| $PR_2$ | $V_2$ | | | | |
| | | First | $AP_{12}$ | $A_{12}$ | $A_{22}–A_{12}$ |
| | | Second | $AP_{22}$ | $A_{22}$ | |
| $PR_3$ | $V_3$ | | | | |
| | | First | $AP_{13}$ | $A_{13}$ | $A_{23}–A_{13}$ |
| | | Second | $AP_{23}$ | $A_{23}$ | |
| $PR_4$ | $V_4$ | | | | |
| | | First | $AP_{14}$ | $A_{14}$ | $A_{24}–A_{14}$ |
| | | Second | $AP_{24}$ | $A_{24}$ | |
| $PR_5$ | $V_5$ | | | | |

mathematical formulation for estimated age is given in Eq. 3.12. This table is defined for the PCI index; however, a similar approach could be used for other indexes such as IRI (notice they increase while PCI decreases across time).

$$A_{ij} = V_j + \frac{(\mathrm{PR}_j - \mathrm{AP}_{ij})}{(\mathrm{PR}_j - \mathrm{PR}_{j+1})} \times (V_{j+1} - V_j) \tag{3.12}$$

$A_{ij}$ is estimated an apparent age for each subgroup observation $i$ ($i = 1$ for the initial year, or 2 for the subsequent time point) in subgroup $j$ (1 to 4 lifespan subgroups in this case) $V_j$ is a variable, $V_1$ is usually zero for PCI, and $\mathrm{PR}_j$ is upper performance (PCI) range for subgroups; 100 is often chosen for the first boundary given the theoretical maximum on PCI (for IRI value of 0.6 to 1.5 would be recommended depending on the local observations' minimum value). Finally, $\mathrm{AP}_{ij}$ is the average PCI for observation $i$ in subgroup $j$. To use Eq. 3.12, a trial and error approach is required to solve each iteration for the $j$th category by first assuming a trial value for ($V_{j+1} > V_j$), then calculating $A_{ij}$ ($i = 1, 2$), and finally checking if ($A_{2j}$–$A_{1j}$) = time interval length (i.e., often a 1-year distance between the initial and subsequent time point) while $\mathrm{PR}_j$ is fixed and $\mathrm{AP}_{ij}$ comes from observation in each subgroup.

Step 4: Adjust the variable age in order to match the age difference to the time interval between the first and second observation (for each interval). This approach is made for all subgroups and next the bests curve would be fitted to the estimated points of (PCI, age).

Step 5: Develop the best-fitted curve for pairs of the average condition and corresponding estimated ages. An easy way would be to test different trendline options (such as Exponential, and Polynomial) in MS-Excel and find the best equation through the coefficient of determination, $R_2$ (Fig. 3.7). For the final

**Table 3.7** Example 3.4 Dataset

| BCI | |
|---|---|
| First Assessment (2019) | Second Assessment (2020) |
| 59 | 56 |
| 84 | 80 |
| 79 | 75 |
| 47 | 44 |
| 26 | 25 |
| 67 | 63 |
| 90 | 84 |
| 56 | 53 |
| 42 | 39 |
| 37 | 35 |
| 28 | 26 |
| 62 | 59 |
| 35 | 33 |
| 95 | 90 |
| 80 | 95 |
| 45 | 90 |
| 30 | 100 |

curve, it could be assumed that the start point is PCI of 100 (or the corresponding locally observed for IRI, for Fatigue Cracking [FC], which will be 0% and 0 inches for rutting) at age zero for a brand-new segment. The assigned curve could be further calibrated by transit agencies either by using available actual age and PCI for some segments or future collected data.

**Example 3.4**
Table 3.7 presents assessment results for some hypothetical steel bridges, which are located in similar harsh weather and can be categorized in the same HG. Let us develop a deterioration model based on this dataset.

***Solution*** This dataset may also cover those bridges that are maintained in this period, thus, in the first step, one should extract those samples that show deterioration ($BCI_{2020} < BCI_{2019}$). Next, some ranges (e.g., 100–70, 69–50, 49–30, and 29–0) should be defined. Following the explained steps, the average BCI in each observation for these ranges will be calculated (Table 3.8). Then, using the proposed calculator (Table 3.6), Table 3.8 can be developed for this example, which results in the deterioration model in Fig. 3.12. The model can be adjusted mathematically to match BCI $= y = 100$ at age $= x = 0$ (Table 3.9).

**Table 3.8** Example 3.4 Solution

| Range | BCI | | BCI | |
|---|---|---|---|---|
| | First Assessment (2019) | Second Assessment (2020) | Average (2019) | Average (2020) |
| 70–100 | 95 | 90 | 87.0 | 82.3 |
| | 84 | 80 | | |
| | 79 | 75 | | |
| | 90 | 84 | | |
| 50–69 | 62 | 59 | 61.0 | 57.8 |
| | 67 | 63 | | |
| | 59 | 56 | | |
| | 56 | 53 | | |
| 30–49 | 42 | 39 | 42.0 | 39.3 |
| | 47 | 44 | | |
| | 37 | 35 | | |
| 0–29 | 28 | 26 | 29.5 | 28.1 |
| | 35 | 33 | | |
| | 26 | 25 | | |

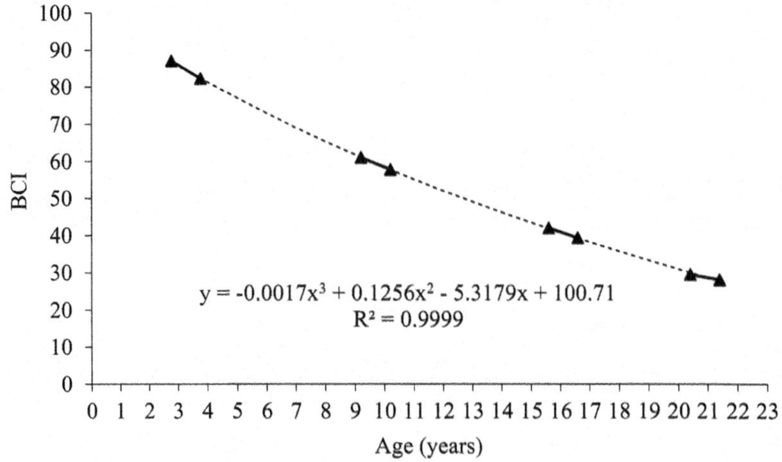


**Fig. 3.12** Example 3.4 deterioration model

#### 3.3.2.2 Stochastic Model (Transition Probability Matrix)

Markov transition probability matrix (TPM) is a widely accepted technique for generating stochastic condition models that reflect the uncertainty of future conditions. The matrix represents the probability distribution of future periods of time. Two cases are possible: (a) a mixed (improvement—deterioration) $TPM_1$ with values all across its cells and (b) a pure deterioration (or improvement) $TPM_2$

**Table 3.9** Hypothetical Example Calculator Results

| Defined Groups Range | Variable Age | Observations | Average BCI | Estimated Age | Age Difference |
|---|---|---|---|---|---|
| 100 | 0.00 | | | | |
| | | First | 87.0 | 2.77 | 1.000 |
| | | Second | 82.3 | 3.77 | |
| 70 | 6.38 | | | | |
| | | First | 61.0 | 9.20 | 1.000 |
| | | Second | 57.8 | 10.20 | |
| 50 | 12.63 | | | | |
| | | First | 42.0 | 15.60 | 1.000 |
| | | Second | 39.3 | 16.60 | |
| 30 | 20.04 | | | | |
| | | First | 29.5 | 20.40 | 1.000 |
| | | Second | 28.1 | 21.40 | |
| 0 | 41.47 | | | | |

composed of zeros below (above) the main diagonal and deterioration (improvement) values above it (below it) (Eq. 3.13). In a mixed TPM, the values above the main diagonal stand for deterioration while the values below it account for treatment improvement. However, for the purpose of AM modeling, a mixed matrix would be confusing where main diagonal probabilities are feasible for both deterioration and improvement and it would be challenging to distinguish. Therefore, to avoid confusion, it would be recommended to develop a separate matrix for deterioration or improvement.

Having two points of condition observation, TPM can be developed while the time interval between two observations (typically one) indicates the prediction period. For instance, in a one-year interval, TPM predicts the next year. Similar to deterministic models, shorter time intervals are recommended to improve the accuracy of a prediction model. In an improvement TPM, two points of data must represent before and after improvement for those that have been maintained.

$$\mathrm{TPM}_1 = \begin{bmatrix} p_{11} & p_{12} & \cdots & p_{1n} \\ p_{21} & p_{22} & \cdots & p_{2n} \\ \vdots & & & \vdots \\ p_{n1} & p_{n2} & \cdots & p_{nn} \end{bmatrix} \quad \mathrm{TPM}_2 = \begin{bmatrix} p_{11} & p_{12} & p_{13} & \cdots & p_{1n} \\ 0 & p_{22} & p_{23} & \cdots & p_{2n} \\ 0 & 0 & p_{33} & \cdots & p_{3n} \\ \vdots & \vdots & \vdots & & \vdots \\ 0 & 0 & 0 & \cdots & 1 \end{bmatrix} \tag{3.13}$$

Each element of the TPM reflects a specific probability. The nomenclature $p_{ij}$ represents the probability that an asset initially in the condition state "$i$" moves to the state "$j$" when a one-time step transition happens (Eq. 3.14). Each row of the TPM matrix sums up to 100% of probable future states (Eq. 3.15). As one moves away from the main diagonal, a higher deterioration or improvement is observed. Many studies have attempted to estimate TPMs. However, the success of this procedure is directly dependent on the existence of quality historical condition data, the precise knowledge of local conditions, and the causal factors, which account for the deterioration in order to divide the information into families of assets (e.g., road segments).

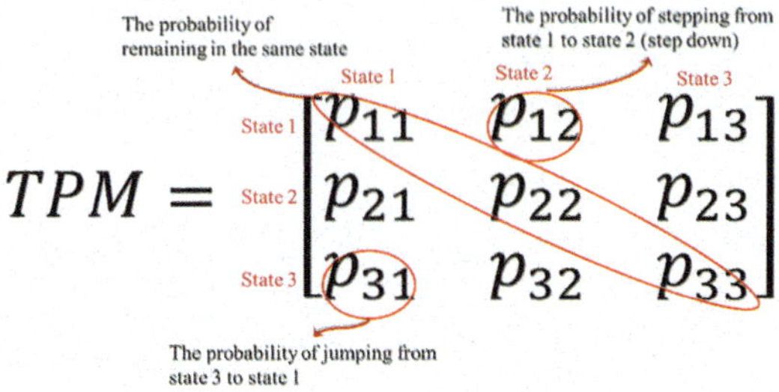


$$p21 + p22 + p23 = 1 \tag{3.15}$$

**Example 3.5**
Fig. 3.13 shows a weather prediction model for city "A". There are two options of Sunny or Rainy. Develop a TPM for this figure.

***Solution*** A stochastic prediction model can be developed for this city using TPM as below:

$$\begin{array}{c} \\ \text{Sunny} \\ \text{Rainy} \end{array} \begin{array}{c} \begin{array}{cc} \text{Sunny} & \text{Rainy} \end{array} \\ \begin{bmatrix} P_{ss} & P_{sr} \\ P_{rs} & P_{rr} \end{bmatrix} \end{array} \Rightarrow \begin{array}{c} \\ \text{Sunny} \\ \text{Rainy} \end{array} \begin{array}{c} \begin{array}{cc} \text{Sunny} & \text{Rainy} \end{array} \\ \begin{bmatrix} 0.8 & 0.2 \\ 0.6 & 0.4 \end{bmatrix} \end{array}$$

To develop a pure deterioration matrix, similar steps as were explained earlier for deterministic can be taken to define homogeneous groups (Step 1) and a separate matrix should be developed for each group.

For each HG, states could be defined by assigning the upper and lower performance boundary (e.g., PCI) and the probability of degrading from state $i$ to the state $j$ ($j > i$) is estimated by counting the total number of segments with primary state of $i$,

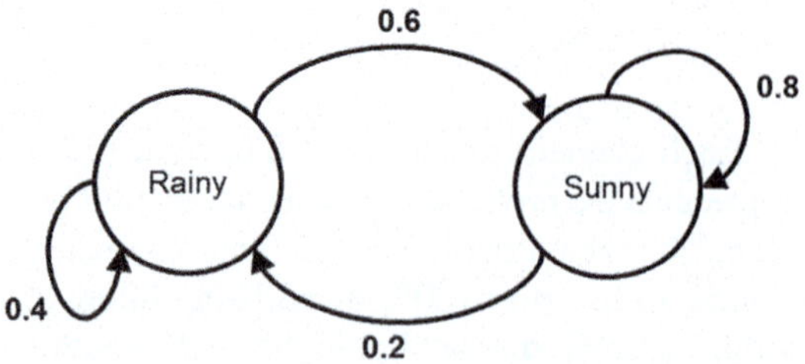


**Fig. 3.13** Example 3.2 dataset

which deteriorated to the state $j$ after the one-time interval. This approach could be presented mathematically as:

$$P_{ij} = \frac{N_{ij}}{N_i} \tag{3.16}$$

Where $P_{ij}$ is the probability of transferring from state $i$ to state $j$ after a time interval, $N_{ij}$ is the number of samples with primary state of $i$ that transferred to state $j$ after a time interval, and $N_i$ shows the total number of samples with a state of $i$ in the first observation.

**Example 3.6**

The small town of "A" has rebuilt the whole 100 road segments last year and, then after 1 year, assessed road conditions while a four-state system (i.e., state 1 presents the best condition) was used by this town for assessment. The results indicate that 70 segments are still in the same state of 1, 20 segments deteriorated to the next state (i.e., state 2), and the rest jumped to state 3. What would be the TPM?

***Solution*** TPM is a matrix of 4*4:

$$TPM = \begin{bmatrix} p_{11} & p_{12} & p_{13} & p_{14} \\ 0 & p_{22} & p_{23} & p_{24} \\ 0 & 0 & p_{33} & p_{34} \\ 0 & 0 & 0 & 1 \end{bmatrix}$$

Since all samples (road segments) were brand new last year (at the beginning of one-year interval), therefore, only probabilities in the first state can be estimated and, for the rest, the town needs to continue assessment in the future. In this example, $N_1 = 100$, and according to the below equations, probabilities can be estimated:

$$p11 = \frac{70}{100} = 0.7$$

$$p13 = \frac{10}{100} = 0.1$$

$$p12 = \frac{20}{100} = 0.2$$

$$p14 = \frac{0}{100} = 0$$

With this observation, the town cannot judge for $p_{22}$, $p_{23}$, $p_{34}$, and $p_{34}$; however, continuing the same exercise will give the chance of populating these probabilities.

**Table 3.10** Example 3.7 Dataset

| | | State I | State II | State III | State IV | |
|---|---|---|---|---|---|---|
| States | PCI Range (%) | 100–70 | 70–50 | 50–30 | 30–0 | Total Number of Segments |
| State I | 100–70 | 750 | 100 | 50 | 0 | 900 |
| State II | 70–50 | 0 | 500 | 300 | 200 | 1000 |
| State III | 50–30 | 0 | 0 | 200 | 600 | 1000 |
| State IV | 30–0 | 0 | 0 | 0 | 1000 | 1000 |

**Table 3.11** Example 3.7 Solution

| | | State I | State II | State III | State IV |
|---|---|---|---|---|---|
| States | PCI range (%) | 100–70 | 70–50 | 50–30 | 30–0 |
| State I | 100–70 | 83% | 11% | 6% | 0% |
| State II | 70–50 | 0% | 50% | 30% | 20% |
| State III | 50–30 | 0% | 0% | 25% | 75% |
| State IV | 30–0 | 0% | 0% | 0% | 100% |

**Example 3.7**
City "B" developed the below table based on the whole city assessment (3900 road segments) in 2018 and 2019. What would be the TPM for roads in this city? Table 3.10 shows the dataset for City B.

***Solution*** In this example, we are able to calculate all probabilities in the states I to IV by the same approach as Example 3.7 (Table 3.11).

**Predict Future by Markov Matrix**
This TPM can be used to predict the future condition of the assets. Condition probability (CP) is a ($1*n$) matrix ($n$ is the number of states in TPM) representing the current condition of the asset:

$$Cp_t = [p1 \;\; p2 \;\; \cdots \;\; pn] \tag{3.17}$$

By multiplying the CP matrix into TPM, the next year's condition of the asset would be presented by an updated CP matrix (Eq. 3.18). CP of newly constructed or installed assets could be estimated for year 1(depends on time interval) as below while $C_{t=0} = [1 \;\; 0 \;\; \cdots \;\; 0]$ is considered for a brand-new asset presented in $J$ states.

$$\begin{aligned} Cp_1 &= C_0 \times P = [1 \;\; 0 \;\; \cdots \;\; 0] \times [TPM] \\ &= [cp_{11} \;\; cp_{12} \;\; \cdots \;\; cp_{1J}] \end{aligned} \tag{3.18}$$

$cp_{ni}$ ($i = 1$ to $k$) is a condition probability for states $i$ in year $n$, $cp_{1j}$ ($j = 1$ to $J$) is a prediction of condition probability for brand new assets after 1 year. For three states matrix, it would be as below:

$$\text{CP} = C_0 \times P = \begin{bmatrix} 1 & 0 & 0 \end{bmatrix} \times \begin{bmatrix} p_{11} & p_{12} & p_{13} \\ 0 & p_{22} & p_{23} \\ 0 & 0 & 1 \end{bmatrix} = \begin{bmatrix} p_{11} & p_{12} & p_{13} \end{bmatrix} \quad (3.19)$$

For older assets, the same approach can be used to predict the future and the current condition of assets ($Cp_t$) should be presented by a matrix and probabilities:

$$Cp_{t+\Delta t} = [Cp_t]_{(1*n)} \times [\text{TPM}]_{(n*n)} \quad (3.20)$$

$$Cp_{t+\Delta t} = \begin{bmatrix} cp_1 & \cdots & cp_n \end{bmatrix} \times \begin{bmatrix} p_{11} & \cdots & p_{1n} \\ 0 & \cdots & \cdots \\ 0 & \cdots & p_{nn} \end{bmatrix} = \begin{bmatrix} cp_{11} & \cdots & cp_{1n} \end{bmatrix} \quad (3.21)$$

$\Delta t$ indicates the time interval for TPM, which typically is 1 year.

Based on a common approach in Markov models, this can be expanded for a future longer time while the prediction matrix for year $n$ ($\text{TPM}_n$) is estimated by:

$$\text{TPM}_n = [\text{TPM}]^n \quad (3.22)$$

Then the whole service life of a new asset can be predicted by Eq. 3.23:

$$\text{CP}_n = C_0 \times \text{TPM}_n = \begin{bmatrix} 1 & 0 & \cdots & 0 \end{bmatrix} \times [\text{TPM}]^n \quad (3.23)$$

$$\text{CP}_n = \begin{bmatrix} cp_{11n} & cp_{12n} & \cdots & cp_{1kn} \end{bmatrix} \; n : 1 \text{ to } N \quad (3.24)$$

Having a probability matrix as the output of the prediction model mathematically makes planning more complicated, and this is possibly the main reason for the less popularity of probabilistic methods among practitioners. A simplified solution would be converting the predicted $\text{CP}_n$ matrix to deterministic results. In Example 3.7, the $\text{CP}_1$ for a brand new road would be:

$$Cp_1 = \begin{bmatrix} 0.83 & 0.11 & 0.06 & 0 \end{bmatrix}$$

Using midpoint for each state (e.g., $85 = \frac{100+70}{2}$ for state I), this matrix can be converted to a single PCI predicted for the next year of a brand new road segment. It shows that the PCI drops from 100 to almost 80 after 1 year.

$$Cp_1 = 0.83 \times 85 + 0.11 \times 60 + 40 \times 0.06 = 79.55$$

**Example 3.8**

Predicting the performance of assets A and B after 5 years using below TPM (one-year interval).

$$\text{TPM} = \begin{bmatrix} 0.75 & 0.20 & 0.05 \\ 0 & 0.50 & 0.50 \\ 0 & 0 & 1 \end{bmatrix}$$

A. Brand new
B. $CP_t = [0.25 \quad 0.70 \quad 0.05]$

***Solution*** For year 2, TPM to power 2 must be calculated:

$$\rightarrow \text{TPM}_2 = \begin{bmatrix} 0.75 & 0.20 & 0.05 \\ 0 & 0.50 & 0.50 \\ 0 & 0 & 1 \end{bmatrix} \times \begin{bmatrix} 0.75 & 0.20 & 0.05 \\ 0 & 0.50 & 0.50 \\ 0 & 0 & 1 \end{bmatrix} =$$

$$\rightarrow \text{TPM}_2 = \begin{bmatrix} 0.5625 & 0.25 & 0.1875 \\ 0 & 0.25 & 0.75 \\ 0 & 0 & 1 \end{bmatrix}$$

The same approach can be used to estimate TPM for years 3 to 5:

$$\rightarrow \text{TPM}_3 = \begin{bmatrix} 0.4219 & 0.2375 & 0.3406 \\ 0 & 0.125 & 0.875 \\ 0 & 0 & 1 \end{bmatrix}$$

$$\rightarrow \text{TPM}_4 = \begin{bmatrix} 0.3164 & 0.2031 & 0.4805 \\ 0 & 0.0625 & 0.9375 \\ 0 & 0 & 1 \end{bmatrix}$$

$$\rightarrow \text{TPM}_5 = \begin{bmatrix} 0.2373 & 0.1648 & 0.5979 \\ 0 & 0.0312 & 0.9688 \\ 0 & 0 & 1 \end{bmatrix}$$

Now, asset condition probabilities for 5 years later can be estimated by multiplying the current condition probability to TPM at year 5.

$$\text{Asset } A : Cp_5 = [1 \quad 0 \quad 0] \times \begin{bmatrix} 0.2373 & 0.1648 & 0.5979 \\ 0 & 0.0312 & 0.9688 \\ 0 & 0 & 1 \end{bmatrix}$$

$$\text{Asset } A = [0.2373 \quad 0.1648 \quad 0.5979]$$

$$\text{Asset } B : Cp_5 = \begin{bmatrix} 0.25 & 0.70 & 0.05 \end{bmatrix} \times \begin{bmatrix} 0.2373 & 0.1648 & 0.5979 \\ 0 & 0.0312 & 0.9688 \\ 0 & 0 & 1 \end{bmatrix}$$

$$\text{Asset } B = \begin{bmatrix} 0.0593 & 0.0630 & 0.8776 \end{bmatrix}$$

By comparing condition probability matrices for assets $A$ and $B$ at year 5, the differences between a brand new and an already aged asset can be explored.

**TPM Limitations**

Besides the popularity and advantages of this method to develop deterioration models for infrastructure, there are some limitations associated with using TPM:

- Asset conditions are presented as a series of discrete states
- A memory_less model
- Ignores age of the asset
- Time homogeneity: TPM remains the same over the whole service life
- Needs time-series data

Alternative probabilistic models, like Semi-Markov (e.g., Markov-Weibull)—a time-based model to enhance Markov's limitations (memory_less & age-independency)—have been recently proposed in the literature (Sobanjo, 2011; Thomas & Sobanjo, 2016).

### *3.3.3 Sufficient Time-Series Data*

The methods presented in this section require time-series data from a sufficient number of years (a minimum of 5), with data characterizing the causal factors (independent variables) and the condition indicator of interest (response or dependent variable). Time-series data comes from multiple observations of the same elements or assets across various years.

Two products are to be estimated: the decay of condition (further explained in this section) and the effectiveness of a given treatment (presented in Chap. 4). Some of the existent methods to estimate the decay of conditions are applicable even in cases when there are missing data elements producing incomplete sequences, which is rather common.

Just as explained in Sect. 3.3, the methods could produce the mean expectation of decay of condition or an enveloping curve for a given confidence interval. The next subsection explains how to obtain either and use some common software to do it.

#### 3.3.3.1 Regression

Regression methods can be applied to estimate how a given condition indicator changes across time. Consider, for example, pavement cracking percentage for a group of HMA arterial roads, all with similar ESALs in a close range between one million and two million, all within the same environmental zone (let us imagine with a moisture index of 80% in a cold region). All that matters from the previous is that, somehow, someone has created HGs and so we can take all observations from the group as belonging to multiple points in the life of the same individual, and use them for the creation of a simple linear or non-linear regression model. As was discussed in Sect. 3.3, a non-linear trend represents a better reality but, in some cases, due to lack of historical data, linear regression also has been used.

There are multiple ways to create a regression, evidently, MS-Excel count with some basic regression forms through the <INSERT> <CHARTS> <SCATTER> menu. Then, the Trendline feature provides a variety of options to fit the best line or curve. Alternatively, the user can create their own named lists and take advantage of the command LINEST to control for the regression form and use named lists to enter the dependent and independent variables. The command INDEX is necessary to extract the desired attribute from the LINEST command. Follow these steps to define a linear regression on Excel:

- Define a named list for the dependent variable, for example, AM_P_WD
- Define a named list for the independent variable, for example, IV
- Use the LINEST command to extract the coefficients of the regression as follows:
- INDEX(LINEST(AM_P_WD,IV,TRUE),1,1): is the slope
- INDEX(LINEST(MD_P_WD,IV,TRUE),1,2): is the intercept
- INDEX(LINEST(MD_P_WD,IV,TRUE,TRUE),3,1): is the coefficient of determination

Time series data is often contained in a large database, and it is always recommended to create a routine (script) to extract the data that you desire according to the HGs for which you want to develop a condition prediction for the decay.

Fig. 3.14 shows an example of using a non-linear regression model for the IRI indicator.

**Example 3.9**
Let us now consider an example of the superstructure of four bridges, all located in the same environmental region, all made out of steel members (superstructure) as shown in Table 3.12. The collected data shows the first 9 years of all four bridges.

***Solution*** To be able to obtain a deterministic curve for these bridges, one must alter the database and stack them vertically using the age (instead of calendar year) information. The best-fit equation is then obtained (Fig. 3.15) and used to obtain discrete ranges for every year of deterioration.

In the previous examples, the data set provided time-series data that represents a pair observation of condition (e.g., IRI) and corresponding age. In some cases, the

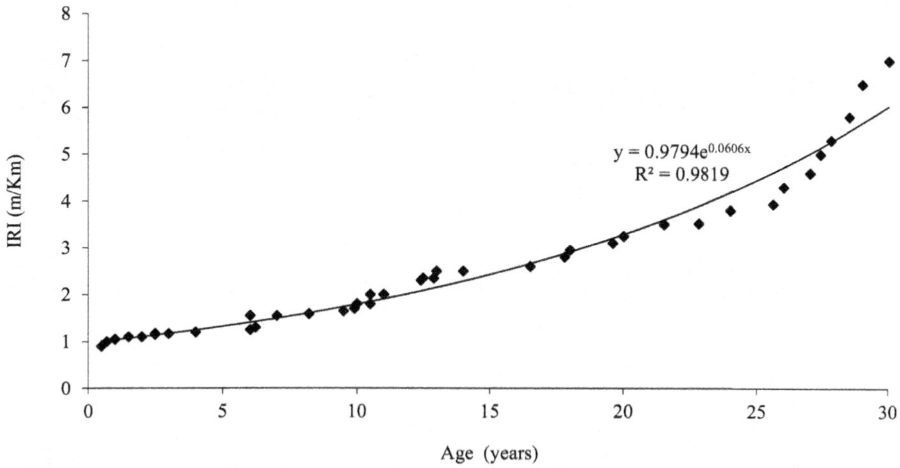


**Fig. 3.14** Non-linear regression model for deterioration

**Table 3.12** Historical Records of Bridge Corrasion for 4 bridges

| Time | Year | Bridge 1 | Year | Bridge 2 | Year | Bridge 3 | Year | Bridge 4 |
|---|---|---|---|---|---|---|---|---|
| 1 | 1991 | 1 | 1985 | 1.5 | 2001 | 0.95 | 1995 | 1.9 |
| 2 | 1992 | 2.25 | 1986 | 2.5 | 2002 | 2.25 | 1996 | 2.78 |
| 3 | 1993 | 5.5 | 1987 | 5.1 | 2003 | 4.8 | 1997 | 6.22 |
| 4 | 1994 | 7.75 | 1988 | 8.13 | 2004 | 7.54 | 1998 | 8.34 |
| 5 | 1995 | 9.88 | 1989 | 10.57 | 2005 | 8.99 | 1999 | 10.85 |
| 6 | 1996 | 12 | 1990 | 12.4 | 2006 | 11.42 | 2000 | 12.79 |
| 7 | 1997 | 15 | 1991 | 14.78 | 2007 | 13.85 | 2001 | 16.55 |
| 8 | 1998 | 18 | 1992 | 18.96 | 2008 | 17.21 | 2002 | 19.1 |
| 9 | 1999 | 21 | 1993 | 22 | 2009 | 20.65 | 2003 | 23 |

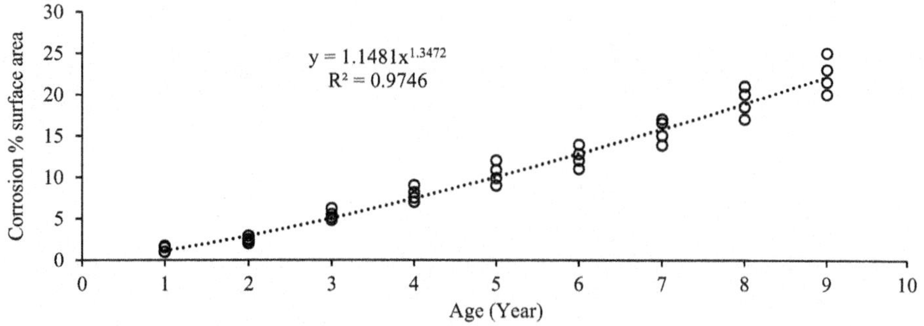


**Fig. 3.15** Deterioration curve extracted from historical observations of corrosion progression

**Table 3.13** Example 3.10 Dataset

| Segment | Year 1 | Year 2 | Year 3 | Year 4 | Year 5 |
|---|---|---|---|---|---|
| 1 | 79.4 | 73.5 | 68.0 | 63.0 | 58.3 |
| 2 | 95.4 | 88.3 | 81.8 | 75.7 | 70.1 |
| 3 | 17.9 | 16.5 | 15.3 | 14.2 | 13.1 |
| 4 | 38.6 | 35.7 | 33.1 | 30.6 | 28.4 |
| 5 | 50.0 | 46.3 | 42.9 | 39.7 | 36.8 |
| 6 | 66.2 | 61.2 | 56.7 | 52.5 | 48.6 |
| 7 | 29.9 | 27.7 | 25.7 | 23.8 | 22.0 |
| 8 | 25.0 | 23.2 | 21.4 | 19.8 | 18.4 |

data set only provides the condition and there is no idea about when exactly the road segment was built or renovated to identify age. A justified method would be used in such a case, which is explained in Example 3.10.

**Example 3.10**
Eight road segments in the same HG are observed for 5 years by assessing PCI (Table 3.13). We want to develop a deterioration model for this dataset.

***Solution*** We have no idea about the age of each segment at the beginning; however, the table shows how PCI drops from each year to the next one. Actually, each row in the table presents the trend of deterioration for a 5-year period. It can be assumed that segment 2, for instance, represents the first beginning 5 years while PCI = 95.4 can represent the road condition at age 1 or segment 3 shows the end of service life. Also, it is a correct assumption that segments with close PCI probably are at the same age meaning that segment 5 for instance at year 1 was at the same/close age as segment 6 after 5 years. (i.e., segment 6 is 5 years younger). This would be the main baseline to develop Table 3.14. Segments 1 to 8 are mapped matching together based on PCI in different years. Next, if we assume that PCI = 95.4 is reflecting age 1, then, the next year would be 2, and the years after is age 3 all to the end. As can be seen for some ages (e.g., age 6 to 8), more than one observation is collected and an average condition can be estimated as the corresponding PCI for these ages. Finally, a bunch of pairs of observations (age and corresponding PCI) are identified, which can be used to project the best-fitted deterioration model (Fig. 3.16).

#### 3.3.3.2 Artificial Intelligent

New technology and clouds give an opportunity to continuously collect data for a longer time period, which means dealing with big data. For instance, thermal condition attributes such as temperature and humidity represent HVAC system performance, and using sensors these attributes can be measured let's say per minute (could be per second depending on asset). Processing and analyzing such an amount of information using classical regression tools would not be practical and the team needs to take advantage of artificial intelligent (AI) to clean and analyze big data.

**Table 3.14** Example 3.10 Solution

| | | | | | | | | | | | | | | | | | | | | | | | | | | |
|---|---|---|---|---|---|---|---|---|---|---|---|---|---|---|---|---|---|---|---|---|---|---|---|---|---|---|
| S 1 | | | | 79 | 73 | 68 | 63 | 58 | | | | | | | | | | | | | | | | | | |
| S 2 | 95 | 88 | 82 | 76 | 70 | | | | | | | | | | | | | | | | | | | | | |
| S 3 | | | | | | | | | | | | | | | | | | | | | | 18 | 17 | 15 | 14 | 13 |
| S 4 | | | | | | | | | | 50 | 46 | 43 | 40 | 37 | | | | | | | | | | | | |
| S 5 | | | | | | | | | | | | | 39 | 36 | 33 | 31 | 28 | | | | | | | | | |
| S 6 | | | | | | 66 | 61 | 57 | 53 | 49 | | | | | | | | | | | | | | | | |
| S 7 | | | | | | | | | | | | | | | | 30 | 28 | 26 | 24 | 22 | | | | | | |
| S 8 | | | | | | | | | | | | | | | | | | 25 | 23 | 21 | 20 | 18 | | | | |
| Age | 1 | 2 | 3 | 4 | 5 | 6 | 7 | 8 | 9 | 10 | 11 | 12 | 13 | 14 | 15 | 16 | 17 | 18 | 19 | 20 | 21 | 22 | 23 | 24 | 25 | 26 |
| PCI | 95 | 88 | 82 | 78 | 72 | 67 | 62 | 58 | 53 | 49 | 46 | 43 | 39 | 36 | 33 | 30 | 28 | 25 | 23 | 22 | 20 | 18 | 17 | 15 | 14 | 13 |

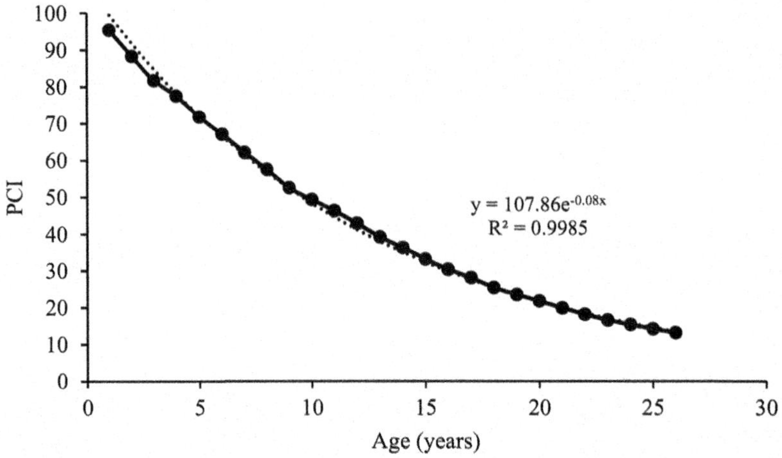


**Fig. 3.16** Example 3.10 solution

The literature is now rich in AI models to find clusters (i.e., homogenous groups) and develop a deterioration model where enough time-series data is available.

## 3.4 Safety and Congestion Performance Prediction

The prediction of performance can be traced back to its ability to explain the accomplishment of the government or the agency in charge of the given asset. Evidently, the most common interest is to count on an asset that exhibits good levels of physical condition as discussed in the section before. However, it is also important to include additional indicators toward applying a performance-based scenario, which relate to the ability of the asset to serve its intended use. Think of a road that is in perfect condition, brand-new like, but that experiences high levels of congestion and observes a high level of collisions, most of them involving serious injuries or fatalities. If the sole goal of the road administrator is a smooth pavement that minimizes damage on vehicles and freight being carried, then the given road compliance is impeccable. However, from a broader perspective, the road is not performing adequately either in terms of mobility or in terms of safety.

Mobility is key to ensuring adequate service of a road because it reduces travel time and facilitates accessibility to land use opportunities (education, health, labor, shops, entertainment, etc.). Good mobility can be captured through the simple relationship of the number of vehicles using each lane divided by the capacity. There are elaborated equations relating current speed with congestion, in simple terms, the lower the level of congestion the faster speed, up to the point of free flow.

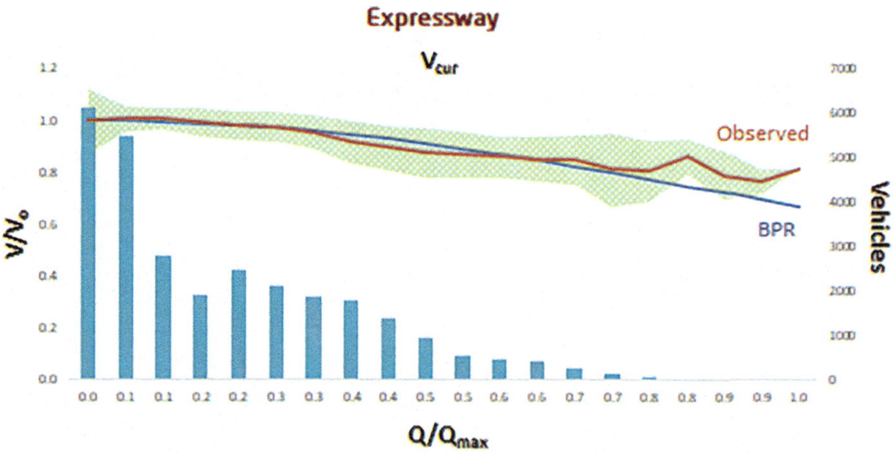


**Fig. 3.17** Sample theoretical (blue line) versus observed (red line) speed—flow relation

$$V_{\text{cur}} = \frac{V_0}{1 + a * \text{sat}^b} \text{ and } sat = \frac{Q}{Q_{\max} * c} \tag{3.25}$$

Where $V_{\text{cur}}$ is current speed (km/h), $V_0 =$ free flow speed $\left(\frac{\text{km}}{\text{h}}\right)$, $Q_{\max} =$ maximum capacity of the link $\left(\frac{\text{PCU}}{\text{hr}}\right)$, $Q =$ current traffic volume (PCU/hr), $V_0 =$ free flow speed$\frac{\text{km}}{\text{hr}}$. Fig. 3.17 illustrates theoretical (blue line) versus observed (red line) speed flow relationships for an expressway.

Safety is a rather different phenomenon, which can be explained by factors such as traffic volume (evidently, the more vehicles, the better odds to experience collisions), speed (the higher the speed, the more severe the consequences of a collision are), and other factors such as the presence of illumination, the presence of safety hardware such as barriers, guardrails, rumble strip), also the geometric characteristics (radius of curvature, the width of lanes and shoulder, separation or not of the carriageway) and the degree of compliance with design standards (sight distance, maximum superelevation, vertical grades change, i.e., k-values, etc.). There are equations for safety called safety performance functions that relate each contributor factor with the total observed number of collisions, in some cases, normalized per kilometer.

Another system that has gained popularity is the road protection scores that mix elements from the following to estimate a score called star rating (Fig. 3.18):

The road protection scores relate to the degree of deficiency of a road and the need to rectify or retrofit the facility. Recently, some attempts have been done to mix both protection scores and safety performance.

Safety performance is important because it can be used to prioritize the correction for deficiencies on roads. For a given type of road (based on its functional classification and traffic volume), one can establish the average number of expected collisions. To such a number, one can subtract the average value of collisions

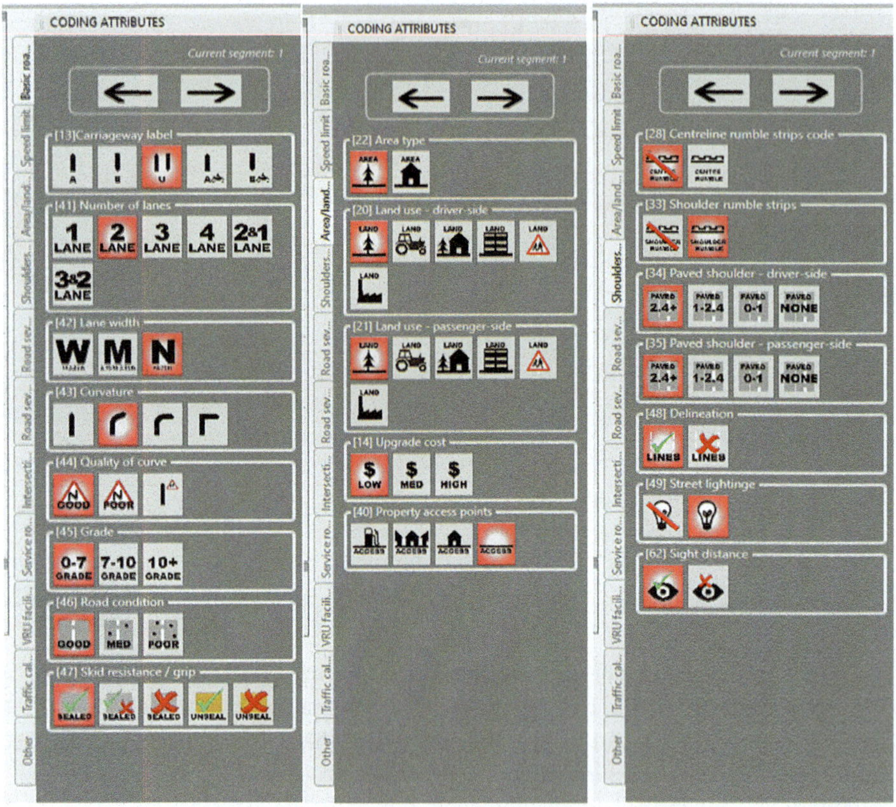


**Fig. 3.18** Sample of iRAP safety coding [S4RD tool—iRAP]

estimated over the past 3–5 years; if such a number is positive, then there are underlying issues that are resulting in an abnormally high number of collisions, and such roads should be scheduled first.

Unfortunately, in the world of road safety, it is common that minor collisions resulting in property damage are not reported, and hundreds of conflicts that result in near misses go silent. For this reason, the use of cameras, marking damage on the road and pavement, and reports of abrupt acceleration/deceleration gathered from insurance companies can be utilized to complement the safety performance to further typify it.

Prediction of safety performance is important to schedule corrective or mitigation measures. Prediction of safety requires the analysis of the before and after trends in terms of the normalized number of collisions, their severity, and the number of conflicts across the road network, with measures being prioritized first to the most severe collisions to reduce fatalities and major injuries, followed by a second wave of measures to counteract minor injuries as well.

Safety extends beyond roads and encompasses pedestrians and cyclists who are vulnerable road users. Safety extends to other assets but takes a slightly different approach in which risk of failure is measured and captured through hazards models, which in turn produce a forecast of the consequences of failure and facilitate the programing of countermeasures to retrofit bridges, and facilities.

## 3.5 Performance Prediction for Demand Capacity

Performance goes beyond the condition of an asset (Sect. 3.3) and its safe operation (Sect. 3.4) and includes the ability to accommodate the demand observed. This section is not intended to replace well-established theories for the estimation of demand and capacity, but rather to bring to the attention of the reader the need to consider them as an integral part of the overall concept of "infrastructure performance" to ensure an asset management system that increases the capacity of the system as a response to permanent increases on the demand originating in urban and economic development.

### *3.5.1 Water Systems*

Water pipes (an important part of water systems) fulfill the role of movement of potable water for human consumption and evacuate wastewater into treatment facilities and stormwater into drainage structures, rivers, and other water bodies. The pipe has a limited hydraulic capacity; in some cases, the pipe can work under pressure (water mains), in others not. Troubles arise when the amount of water that rains on a given catchment area of a storm pipe are more than the amount the pipe is designed for. In this eventuality, the drains cannot take water as fast as it arrives, creating standing water and flooding of the immediate urban area around. For a water main delivering water at multiple uses of land (residential, commercial, industrial, etc.), not having enough water—either being fed in real-time at the water treatment plant or holding on into tanks to cover the peak demand periods—is an issue.

Demand for rain can be tied naturally with climate change and weather cycles; models such as the rational method can provide simple mechanisms to estimate the demand of water storm systems, controlling for intensity, the period of return of rain, but also for the size of the catchment area and the land use development through the runoff coefficients. Chap. 6 presents an example of how demand capacity can be addressed in AM decision-making for stormwater pipelines addressing climate change.

### 3.5.2 *Transportation*

There are important interdependencies between the demand, capacity, and service quality of the public transit system. Multiple factors contribute to the demand capacity ratio in a transit system. Forecasting demand is a complex task. It requires a clear understanding of present and future land use in the area, and the demand each land use will impose. Capacity also shows the ability of a system to handle commuters (e.g., number of passenger cars that the road lane can handle per hour). Appendix 2 provides a high-level discussion around demand and capacity predictions; however, the literature is rich for these topics, and transit agencies depending on the size and scale of their system need to implement the demand-capacity study.

Higher future demand and ridership mean faster deterioration due to increasing traffic load in roads or rail tracks and a higher usage rate in station facilities. Thus, AM needs to address this change for existing assets. On the other hand, a perfect SGR plan delivers improved ride quality (e.g., vehicle comfort), which leads to increased demand and ridership (Mohammadi et al., 2020). Moreover, by upkeeping assets in the highest availability, AM directly contributes to system capacity while higher capacity results in more complex planning for AM.

Decision-makers and asset managers cannot ignore these dependencies and need to closely collaborate with the demand planning team and reflect it in long-term planning.

## 3.6 Exercise

**Exercise 3.1**

Estimate the decay of condition in 15 years of a pavement using the simplified deterministic method proposed by Patterson and Attoh-Okine (shown below) in which NESALs are in millions, m is Thornthwaite's moisture index, defined through the matching of moisture index zones, IRI0 is initial IRI and SNC is the structural number which is estimated from the pavement layers thickness ($H_{\text{ac}} = 6$ and $H_{\text{Base}} = 10$ inches) and structural coefficients as recommended by AASHTO 1993; in this case, assumed as 0.35 for the hot mix asphalt layer and 0.15 for the granular base. Consider 0.75 million ESALs (load repetitions of 9000 lbs. on circular contact area from the imprint of dual tires on a radius between 4 and 6 inches). Traffic growth is 3%.

$$\text{IRI}_t = e^{0.07t}\left[\text{IRI}_0 + 250 * \frac{NE_t}{(1+\text{SNC})^5}\right]$$

$$\text{SNC} = 0.35 * H_{\text{AC}} + 0.15 * H_{\text{Base}}$$

**Exercise 3.2**

Estimate the decay of condition for bottom-up FC (illustration below) for the same asphalt layer of 6 inches of Exercise 3.1; elastic modulus is 100,000 psi, Poisson ratio is 0.35, the load is $P = 9000$ lb (1 ESAL) on an area of 63.6172 in$^2$ (radius 4.5), tensile strain at the bottom of the asphalt layer is approximately estimated by the following equation. To generate the deterioration curve, use various values of observed loads: million, two million, three million,...14 million. Assume asphalt binder volume ($V_b$) = 0.11, and air volume in the mix ($V_a$) = 0.05 (Fig. 3.19).

The model will be based on the mechanistic-empirical pavement design guide (AASHTO, 2008) for FC and FD which are instrumented in the following equations.

$$\mathrm{FC} = \frac{1560}{1 + e^{(7-3.5\log \mathrm{FD})}}$$

$$\mathrm{FD} = \sum \frac{n_{i,j,k,l,m}}{N_{i,j,k,l,m}}, \text{ simplifies in this case } = \frac{n_{9000 \text{ lbs load at 85 Fahrenheit}}}{N_{9000 \text{ lbs load at 85 Fahrenheit}}}$$

Where

$$N_f = 0.00432 \cdot k_1' C \left(\frac{1}{\varepsilon_t}\right)^{3.9492} \left(\frac{1}{E}\right)^{1.281}$$

$$k_1' = \frac{1}{0.000398 + \frac{0.003602}{1+e^{11.02-3.49 h_{ac}}}}$$

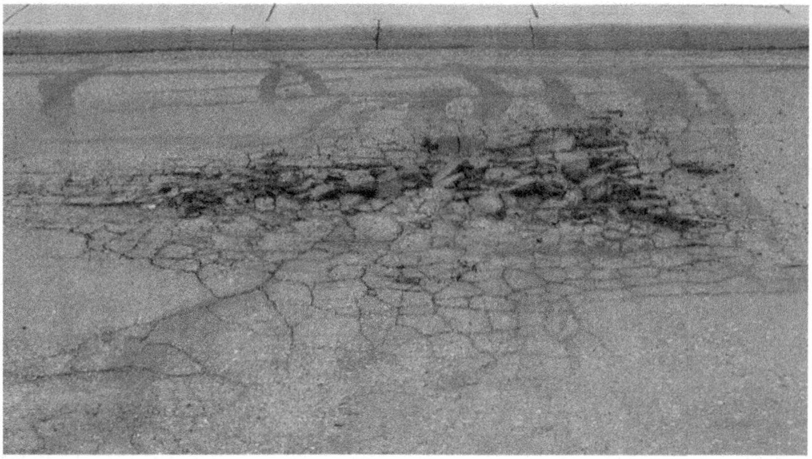

Fig. 3.19 Exercise 3.2 Dataset

$$C = 10^{4.84\left(\frac{V_b}{V_a+V_b}-0.69\right)}$$

**Exercise 3.3**

Estimate the decay of condition for plastic deformations (illustration below) for the same asphalt layer of 6 inches of Example 1; elastic modulus is 100,000 psi, Poisson ratio is 0.35, the load is $P = 9000$ lb. (1 ESAL) on an area of 63.6172 in$^2$ (radius 4.5), vertical elastic strain at the bottom of the asphalt layer is approximately estimated by the following equation. To generate the deterioration curve, use various values of observed loads: one million, two million, three million,. . .12 million. Assume asphalt binder volume ($V_b$) = 0.11, and air volume in the mix ($V_a$) = 0.05.

PD $= \varepsilon_p * h$, where : $\varepsilon_p$ = plastic strain in AC layer and $h$ = thickness of AC layer.

$\varepsilon_p = \varepsilon_v * 1.654 * 10^{-3.4488} T^{1.5606} N^{0.479244}$, $T$ = temperature in Fahrenheit, $N$ = number of load repetitions

$$\varepsilon_v = \frac{(1+\upsilon)q}{E}\left[1 - 2\upsilon + \frac{2\upsilon z}{(a^2+z^2)^{0.5}} - \frac{z^3}{(a^2+z^2)^{1.5}}\right]$$

**Exercise 3.4**

Create a historical deterioration model for a concrete bridge deck that suffers from delamination. Delamination is produced by horizontal cracking caused by corrosion of the embedded reinforcing steel. The bridge deck has been in constant monitoring for the past 25 years through laser measurements on the bottom face of the deck, and the values observed for a group of 10 bridges all on the same highways and of the similar span are presented in Table 3.15.

**Exercise 3.5**

Calibrate a safety performance curve using historical measurements of causal factors:

- Volume: Average number of vehicles per day per lane
- Speed: Standard deviation for the 85 percentile operational speeds/100 (in km/h)
- Driveways: Number of driveways per kilometre

Use the classical form: Road Collisions = POWER (Volume, beta1) *EXP (Speed*beta2 + driveways*beta3)

We will depart from the following data (typically the dataset will be much larger)

**Table 3.15** Exercise 3.4 Dataset

| Year | Deck_1 | Deck_2 | Deck_3 | Deck_4 | Deck_5 | Deck_6 | Deck_7 | Deck_8 | Deck_9 | Deck_10 |
|---|---|---|---|---|---|---|---|---|---|---|
| 1 | 0.040 | 0.020 | 0.034 | 0.030 | 0.010 | 0.030 | 0.0125 | 0.037 | 0.0245 | 0.0126 |
| 2 | 0.035 | 0.035 | 0.039 | 0.015 | 0.005 | 0.025 | 0.0175 | 0.0285 | 0.0585 | 0.0178 |
| 3 | 0.050 | 0.050 | 0.037 | 0.030 | 0.040 | 0.040 | 0.0225 | 0.047 | 0.049 | 0.0229 |
| 4 | 0.055 | 0.055 | 0.054 | 0.035 | 0.045 | 0.035 | 0.0325 | 0.0655 | 0.054 | 0.033 |
| 5 | 0.060 | 0.050 | 0.045 | 0.040 | 0.040 | 0.030 | 0.0375 | 0.0435 | 0.0735 | 0.0381 |
| 6 | 0.065 | 0.055 | 0.052 | 0.045 | 0.055 | 0.045 | 0.0375 | 0.0485 | 0.064 | 0.03785 |
| 7 | 0.073 | 0.063 | 0.072 | 0.043 | 0.043 | 0.043 | 0.05 | 0.083 | 0.086 | 0.051275 |
| 8 | 0.080 | 0.070 | 0.107 | 0.020 | 0.070 | 0.050 | 0.0625 | 0.0635 | 0.137 | 0.0645 |
| 9 | 0.088 | 0.068 | 0.093 | 0.078 | 0.078 | 0.038 | 0.07 | 0.1385 | 0.1445 | 0.0723 |
| 10 | 0.125 | 0.115 | 0.118 | 0.085 | 0.065 | 0.065 | 0.0725 | 0.0785 | 0.1085 | 0.0764 |
| 11 | 0.133 | 0.133 | 0.112 | 0.123 | 0.123 | 0.073 | 0.08 | 0.113 | 0.145 | 0.08435 |
| 12 | 0.110 | 0.140 | 0.100 | 0.090 | 0.050 | 0.060 | 0.09 | 0.107 | 0.1235 | 0.0932 |
| 13 | 0.118 | 0.118 | 0.130 | 0.098 | 0.108 | 0.078 | 0.0925 | 0.1145 | 0.131 | 0.095125 |
| 14 | 0.105 | 0.115 | 0.123 | 0.085 | 0.075 | 0.095 | 0.105 | 0.1085 | 0.124 | 0.1088 |
| 15 | 0.113 | 0.123 | 0.125 | 0.053 | 0.083 | 0.093 | 0.1125 | 0.1565 | 0.175 | 0.117625 |
| 16 | 0.170 | 0.150 | 0.129 | 0.110 | 0.150 | 0.130 | 0.12 | 0.1775 | 0.1245 | 0.1222 |
| 17 | 0.183 | 0.223 | 0.179 | 0.143 | 0.093 | 0.093 | 0.13 | 0.1495 | 0.1515 | 0.13735 |
| 18 | 0.235 | 0.215 | 0.222 | 0.175 | 0.175 | 0.155 | 0.155 | 0.2295 | 0.1785 | 0.15905 |
| 19 | 0.218 | 0.208 | 0.257 | 0.198 | 0.138 | 0.138 | 0.1725 | 0.2285 | 0.191 | 0.18725 |
| 20 | 0.190 | 0.210 | 0.208 | 0.140 | 0.120 | 0.160 | 0.1675 | 0.295 | 0.218 | 0.1803 |
| 21 | 0.243 | 0.243 | 0.219 | 0.223 | 0.213 | 0.193 | 0.195 | 0.24 | 0.2305 | 0.20535 |
| 22 | 0.228 | 0.198 | 0.250 | 0.178 | 0.188 | 0.208 | 0.1925 | 0.3225 | 0.318 | 0.2075 |
| 23 | 0.253 | 0.243 | 0.292 | 0.153 | 0.203 | 0.173 | 0.2275 | 0.243 | 0.275 | 0.2356 |
| 24 | 0.248 | 0.308 | 0.299 | 0.168 | 0.158 | 0.168 | 0.2425 | 0.339 | 0.3045 | 0.26425 |
| 25 | 0.243 | 0.263 | 0.275 | 0.183 | 0.213 | 0.173 | 0.245 | 0.2865 | 0.305 | 0.2543 |
| 26 | 0.338 | 0.338 | 0.277 | 0.268 | 0.248 | 0.308 | 0.265 | 0.3555 | 0.3635 | 0.26995 |

| Volume | Speed (km/h) | Driveways | Crash |
|---|---|---|---|
| 1.85 | 0.28 | 10 | 12 |
| 1.84 | 0.27 | 18 | 23 |
| 1.80 | 0.32 | 0 | 5 |
| 1.72 | 0.33 | 4 | 7 |
| 1.74 | 0.23 | 2 | 6 |
| 1.89 | 0.2 | 13 | 14 |
| 1.89 | 0.32 | 0 | 6 |
| 1.77 | 0.12 | 3 | 5 |
| 1.85 | 0.32 | 15 | 28 |
| 1.84 | 0.19 | 19 | 34 |
| 1.80 | 0.22 | 10 | 10 |
| 1.72 | 0.35 | 9 | 11 |
| 1.74 | 0.28 | 8 | 9 |
| 1.89 | 0.22 | 1 | 5 |
| 1.89 | 0.31 | 20 | 32 |
| 1.77 | 0.14 | 4 | 5 |
| 1.38 | 0.05 | 16 | 9 |
| 1.85 | 0.28 | 14 | 18 |
| 1.65 | 0.27 | 15 | 16 |
| 1.74 | 0.14 | 17 | 13 |
| 1.81 | 0.09 | 0 | 4 |
| 1.81 | 0.3 | 17 | 28 |
| 1.75 | 0.31 | 12 | 15 |
| 1.84 | 0.11 | 0 | 4 |
| 1.81 | 0.34 | 11 | 15 |
| 1.85 | 0.26 | 7 | 11 |
| 1.65 | 0.19 | 1 | 3 |
| 1.74 | 0.22 | 0 | 4 |
| 1.81 | 0.29 | 9 | 13 |
| 1.81 | 0.09 | 15 | 14 |
| 1.75 | 0.06 | 8 | 6 |
| 1.84 | 0.13 | 5 | 7 |
| 1.81 | 0.08 | 7 | 7 |
| 1.38 | 0.18 | 3 | 3 |

**Exercise 3.6**

Develop a deterioration curve based on collected data for 5 years of observations of pavement condition (PCI) in a small town in Ontario (Table 3.16).

**Exercise 3.7**

Use the Exercise 3.6 dataset and develop Markov transition matrix. Define these states based on PCI: 100-70, 70-50, 50-30, 30-0.

**Table 3.16** Exercise 3.6 Dataset

| Segment | Year 1 | Year 2 | Year 3 | Year 4 | Year 5 |
|---|---|---|---|---|---|
| s1 | 17.2 | 16.3 | 15.4 | 14.6 | 13.8 |
| s2 | 58.3 | 55.2 | 52.2 | 49.4 | 46.8 |
| s3 | 99.0 | 93.7 | 88.7 | 83.9 | 79.5 |
| s4 | 30.9 | 29.2 | 27.7 | 26.2 | 24.8 |
| s5 | 20.7 | 19.6 | 18.5 | 17.6 | 16.6 |
| s6 | 11.7 | 11.1 | 10.5 | 9.9 | 9.4 |
| s7 | 46.3 | 43.8 | 41.5 | 39.3 | 37.2 |
| s8 | 70.5 | 66.7 | 63.1 | 59.7 | 56.5 |
| s9 | 90.5 | 85.6 | 81.1 | 76.7 | 72.6 |
| s10 | 25.0 | 23.7 | 22.4 | 21.2 | 20.1 |
| s11 | 45.4 | 43.0 | 40.7 | 38.5 | 36.4 |
| s12 | 15.6 | 14.7 | 13.9 | 13.2 | 12.5 |
| s13 | 27.3 | 25.8 | 24.4 | 23.1 | 21.9 |
| s14 | 31.5 | 29.8 | 28.2 | 26.7 | 25.3 |
| s15 | 77.5 | 73.3 | 69.4 | 65.7 | 62.2 |
| s16 | 27.0 | 25.5 | 24.2 | 22.9 | 21.7 |
| s17 | 51.2 | 48.4 | 45.8 | 43.4 | 41.1 |
| s18 | 68.1 | 64.4 | 61.0 | 57.7 | 54.6 |
| s19 | 18.8 | 17.8 | 16.9 | 16.0 | 15.1 |
| s20 | 13.5 | 12.8 | 12.1 | 11.5 | 10.9 |
| s21 | 41.9 | 39.7 | 37.5 | 35.5 | 33.6 |
| s22 | 75.2 | 71.2 | 67.4 | 63.8 | 60.4 |
| s23 | 49.9 | 47.2 | 44.7 | 42.3 | 40.0 |
| s24 | 22.7 | 21.4 | 20.3 | 19.2 | 18.2 |
| s25 | 18.0 | 17.0 | 16.1 | 15.3 | 14.4 |
| s26 | 12.9 | 12.3 | 11.6 | 11.0 | 10.4 |
| s27 | 36.8 | 34.8 | 33.0 | 31.2 | 29.5 |
| s28 | 37.9 | 35.9 | 34.0 | 32.1 | 30.4 |
| s29 | 20.5 | 19.4 | 18.4 | 17.4 | 16.5 |
| s30 | 14.1 | 13.3 | 12.6 | 11.9 | 11.3 |

**Exercise 3.8**

Calibrate a demand capacity model for multilane highways in Table 3.17.

**Exercise 3.9**

Calibrate road safety performance function presented below for the following data.

**Table 3.17** Exercise 3.8 Dataset

| Segment_ID | Length (m) | Lanes | $V_0$ | $Q$ (pc/ph/pl) | $V_{cur}$ | $Q_{max}$ |
|---|---|---|---|---|---|---|
| 19,090 | 881 | 7 | 100 | 6285 | 45 | 1524 |
| 19,003 | 3427 | 6 | 100 | 14,568 | 31 | 1717 |
| 19,098 | 1543 | 6 | 100 | 13,708 | 29 | 1408 |
| 19,103 | 1133 | 6 | 100 | 12,927 | 28 | 1239 |
| 20,000 | 3072 | 6 | 100 | 12,797 | 33 | 1722 |
| 40,710 | 3061 | 6 | 100 | 11,312 | 36 | 1754 |
| 40,040 | 3407 | 6 | 100 | 11,258 | 36 | 1780 |
| 19,100 | 1152 | 6 | 100 | 11,032 | 32 | 1361 |
| 19,099 | 725 | 6 | 100 | 9868 | 30 | 1093 |
| 19,097 | 495 | 6 | 100 | 9023 | 26 | 784 |
| 19,095 | 1179 | 6 | 100 | 7998 | 40 | 1548 |
| 19,101 | 1252 | 6 | 100 | 7909 | 41 | 1579 |
| 19,096 | 796 | 6 | 100 | 7238 | 40 | 1394 |
| 19,102 | 757 | 6 | 100 | 6416 | 43 | 1435 |
| 19,089 | 725 | 6 | 100 | 6385 | 43 | 1413 |
| 19,105 | 457 | 6 | 100 | 6371 | 38 | 1072 |
| 19,104 | 463 | 6 | 100 | 5806 | 41 | 1164 |
| 19,010 | 3208 | 5 | 100 | 16,421 | 29 | 1659 |
| 10,070 | 3881 | 5 | 100 | 15,100 | 31 | 1741 |
| 19,002 | 2213 | 5 | 100 | 13,794 | 31 | 1585 |
| 19,024 | 2787 | 5 | 100 | 9043 | 41 | 1784 |
| 30,101 | 2030 | 5 | 100 | 8587 | 41 | 1718 |
| 19,011 | 1265 | 5 | 100 | 6687 | 45 | 1648 |
| 19,001 | 2190 | 5 | 100 | 5777 | 50 | 1824 |
| 10,003 | 7892 | 5 | 100 | 1961 | 59 | 1000 |
| 19,094 | 973 | 4 | 90 | 21,063 | 11 | 557 |
| 10,080 | 6240 | 4 | 90 | 16,150 | 27 | 1827 |
| 19,093 | 617 | 4 | 90 | 14,391 | 12 | 445 |
| 19,038 | 3040 | 4 | 90 | 13,016 | 30 | 1715 |
| 19,026 | 1426 | 4 | 90 | 12,948 | 27 | 1395 |
| 19,068 | 1343 | 4 | 90 | 11,618 | 29 | 1423 |
| 19,006 | 2257 | 4 | 90 | 11,515 | 31 | 1660 |
| 19,063 | 1180 | 4 | 90 | 11,308 | 28 | 1361 |
| 19,023 | 1498 | 4 | 90 | 10,748 | 31 | 1522 |
| 30,600 | 5116 | 4 | 90 | 10,704 | 34 | 1861 |
| 19,004 | 1979 | 4 | 90 | 10,544 | 33 | 1645 |
| 19,062 | 2290 | 4 | 90 | 10,479 | 33 | 1695 |
| 20,010 | 12,087 | 4 | 90 | 10,200 | 36 | 1944 |
| 19,064 | 2295 | 4 | 90 | 10,138 | 34 | 1706 |
| 19,056 | 2829 | 4 | 90 | 10,047 | 34 | 1763 |
| 19,067 | 1637 | 4 | 90 | 9861 | 33 | 1598 |
| 19,005 | 2058 | 4 | 90 | 9754 | 34 | 1684 |

(continued)

**Table 3.17** (continued)

| Segment_ID | Length (m) | Lanes | $V_0$ | $Q$ (pc/ph/pl) | $V_{cur}$ | $Q_{max}$ |
|---|---|---|---|---|---|---|
| 19,092 | 588 | 4 | 90 | 9212 | 26 | 956 |
| 19,017 | 1656 | 4 | 90 | 9053 | 35 | 1636 |
| 19,053 | 961 | 4 | 90 | 8971 | 32 | 1378 |
| 19,012 | 1846 | 4 | 90 | 8620 | 36 | 1689 |
| 19,020 | 773 | 4 | 90 | 8283 | 32 | 1286 |
| 19,037 | 359 | 4 | 90 | 8100 | 19 | 494 |
| 30,061 | 1519 | 4 | 90 | 8032 | 37 | 1647 |
| 19,052 | 1350 | 4 | 90 | 7875 | 37 | 1611 |
| 19,059 | 3109 | 4 | 90 | 7596 | 40 | 1837 |
| 20,250 | 1693 | 4 | 90 | 7376 | 39 | 1709 |
| 19,030 | 1604 | 4 | 90 | 7219 | 40 | 1700 |
| 19,047 | 1096 | 4 | 90 | 6730 | 40 | 1591 |
| 19,070 | 1313 | 4 | 90 | 6673 | 41 | 1661 |
| 19,065 | 3613 | 4 | 90 | 6661 | 43 | 1877 |
| 19,061 | 1661 | 4 | 90 | 6585 | 42 | 1736 |
| 19,050 | 1052 | 4 | 90 | 6400 | 41 | 1595 |
| 19,032 | 1104 | 4 | 90 | 6260 | 41 | 1622 |
| 19,091 | 870 | 4 | 90 | 6010 | 41 | 1540 |
| 30,040 | 2979 | 4 | 90 | 5413 | 47 | 1879 |
| 19,128 | 473 | 4 | 90 | 5242 | 40 | 1261 |
| 19,054 | 2110 | 4 | 90 | 5106 | 47 | 1839 |
| 19,039 | 2917 | 4 | 90 | 4949 | 48 | 1887 |
| 19,058 | 1512 | 4 | 90 | 4861 | 48 | 1786 |
| 40,750 | 2783 | 4 | 90 | 4838 | 49 | 1884 |
| 40,010 | 3617 | 4 | 90 | 4747 | 49 | 1912 |
| 60,621 | 3755 | 4 | 90 | 3632 | 54 | 1936 |
| 10,390 | 2916 | 4 | 90 | 3525 | 55 | 1919 |
| 19,025 | 4150 | 4 | 90 | 3100 | 57 | 1950 |
| 19,112 | 1437 | 4 | 90 | 2782 | 58 | 1871 |
| 40,610 | 421 | 4 | 90 | 2001 | 62 | 1683 |
| 70,410 | 2771 | 4 | 90 | 1396 | 67 | 1670 |
| 51,081 | 973 | 4 | 90 | 48 | 86 | 600 |
| 60,271 | 15,110 | 4 | 90 | 27 | 87 | 600 |
| 19,016 | 1137 | 3 | 80 | 11,584 | 24 | 1321 |
| 40,000 | 4820 | 3 | 80 | 11,083 | 30 | 1847 |
| 10,450 | 2487 | 3 | 80 | 10,520 | 30 | 1718 |
| 19,019 | 1748 | 3 | 80 | 10,210 | 29 | 1611 |
| 19,015 | 355 | 3 | 80 | 9690 | 7 | 179 |
| 10,211 | 1348 | 3 | 80 | 8188 | 32 | 1595 |
| 20,020 | 7062 | 3 | 80 | 7596 | 36 | 1928 |
| 20,031 | 7981 | 3 | 80 | 7478 | 37 | 1938 |
| 19,034 | 2508 | 3 | 80 | 6731 | 37 | 1821 |

(continued)

**Table 3.17** (continued)

| Segment_ID | Length (m) | Lanes | $V_0$ | $Q$ (pc/ph/pl) | $V_{cur}$ | $Q_{max}$ |
|---|---|---|---|---|---|---|
| 19,076 | 548 | 3 | 80 | 6677 | 31 | 1188 |
| 20,132 | 2242 | 3 | 80 | 6576 | 38 | 1804 |
| 20,131 | 2537 | 3 | 80 | 6450 | 38 | 1831 |
| 20,050 | 4773 | 3 | 80 | 6262 | 39 | 1913 |
| 20,040 | 4242 | 3 | 80 | 6247 | 39 | 1902 |
| 19,121 | 788 | 3 | 80 | 5707 | 37 | 1517 |
| 21,430 | 14,001 | 3 | 80 | 5664 | 36 | 1400 |
| 21,890 | 15,215 | 3 | 80 | 5486 | 36 | 1400 |
| 40,180 | 1178 | 3 | 80 | 5133 | 41 | 1709 |
| 10,920 | 1792 | 3 | 80 | 5099 | 42 | 1810 |
| 60,623 | 3812 | 3 | 80 | 4843 | 43 | 1915 |
| 60,180 | 4900 | 3 | 80 | 3919 | 47 | 1947 |
| 21,442 | 6822 | 3 | 80 | 3823 | 48 | 2000 |
| 19,048 | 3259 | 3 | 80 | 3285 | 50 | 1933 |
| 70,160 | 10,274 | 3 | 80 | 2797 | 42 | 1000 |
| 10,950 | 5012 | 3 | 80 | 2790 | 53 | 2000 |
| 21,441 | 11,639 | 3 | 80 | 1711 | 42 | 600 |
| 70,730 | 1932 | 3 | 80 | 32 | 77 | 600 |

$$\text{Crash} = b1^{*}\ln(\text{PC\_MI\_LN\_HR}) + b2^{*}\text{SpeedV85}$$

| Segment | Route | Q_lane | PC_MI _LN_HR | Speed V85 | Crash_ frequency per_km | Crash_ severity per_km | lnPC_MI _LN_HR |
|---|---|---|---|---|---|---|---|
| 19,089 | 39 | 6385.2 | 587.4 | 42.8 | 89.7 | 139.4 | 6.4 |
| 60,210 | 1 | 8729.0 | 129.8 | 17.2 | 66.4 | 116.6 | 4.9 |
| 60,220 | 1 | 5092.5 | 69.9 | 21.4 | 49.8 | 96.2 | 4.2 |
| 20,131 | 3 | 6450.3 | 169.5 | 38.2 | 47.3 | 68.6 | 5.1 |
| 60,230 | 1 | 4960.5 | 30.0 | 25.3 | 44.2 | 112.8 | 3.4 |
| 30,472 | 10 | 2166.0 | 33.9 | 22.8 | 41.8 | 79.9 | 3.5 |
| 19,091 | 39 | 6010.3 | 460.4 | 41.1 | 40.2 | 71.2 | 6.1 |
| 19,023 | 104 | 10747.8 | 478.2 | 30.9 | 31.4 | 31.4 | 6.2 |
| 30,010 | 10 | 1931.0 | 17.2 | 23.8 | 26.4 | 48.1 | 2.8 |
| 10,001 | 2 | 1850.0 | 6.9 | 30.3 | 26.3 | 38.4 | 1.9 |
| 19,017 | 100 | 9053.3 | 364.4 | 35.0 | 22.9 | 39.2 | 5.9 |
| 20,812 | 135 | 7090.5 | 336.3 | 13.9 | 19.2 | 38.4 | 5.8 |
| 19,090 | 39 | 6285.0 | 475.6 | 44.6 | 18.2 | 28.4 | 6.2 |
| 19,008 | 5 | 7279.0 | 211.9 | 13.7 | 17.9 | 41.5 | 5.4 |
| 20,132 | 3 | 6576.3 | 195.5 | 37.6 | 17.8 | 29.9 | 5.3 |
| 10,881 | 326 | 468.5 | 4.5 | 36.4 | 17.6 | 27.9 | 1.5 |
| 30,120 | 251 | 6173.0 | 84.8 | 14.8 | 15.3 | 37.5 | 4.4 |
| 20,200 | 111 | 2957.5 | 100.4 | 20.2 | 13.8 | 18.3 | 4.6 |

| | | | | | | | |
|---|---|---|---|---|---|---|---|
| 40,032 | 106 | 7782.5 | 153.3 | 13.3 | 13.0 | 21.0 | 5.0 |
| 60,052 | 2 | 833.5 | 2.3 | 43.5 | 12.4 | 26.1 | 0.8 |
| 40,031 | 106 | 7165.5 | 191.3 | 13.8 | 12.0 | 22.8 | 5.3 |
| 30,051 | 236 | 5781.5 | 218.6 | 15.2 | 11.9 | 11.9 | 5.4 |
| 40,270 | 220 | 1158.5 | 27.3 | 28.3 | 11.7 | 27.6 | 3.3 |
| 19,062 | 215 | 10479.0 | 305.0 | 33.1 | 10.9 | 30.6 | 5.7 |
| 21,022 | 4 | 932.5 | 4.9 | 46.2 | 10.8 | 28.6 | 1.6 |
| 40,420 | 117 | 2605.5 | 190.5 | 21.3 | 9.9 | 29.6 | 5.2 |
| 20,262 | 124 | 5751.0 | 84.0 | 15.3 | 9.6 | 23.4 | 4.4 |
| 40,500 | 32 | 16469.5 | 144.3 | 15.1 | 9.6 | 26.2 | 5.0 |
| 20,580 | 126 | 668.5 | 4.1 | 33.2 | 9.4 | 22.7 | 1.4 |
| 19,030 | 108 | 7218.8 | 300.0 | 39.6 | 9.4 | 9.4 | 5.7 |
| 70,490 | 32 | 8145.0 | 122.3 | 24.3 | 8.8 | 29.1 | 4.8 |
| 20,261 | 124 | 7496.5 | 186.8 | 13.5 | 8.6 | 18.7 | 5.2 |
| 60,240 | 1 | 4456.5 | 15.5 | 28.9 | 8.4 | 21.6 | 2.7 |
| 60,060 | 2 | 1244.5 | 2.8 | 39.5 | 8.1 | 24.7 | 1.0 |
| 70,150 | 32 | 4922.5 | 24.8 | 25.3 | 7.6 | 19.2 | 3.2 |
| 21,570 | 140 | 1883.5 | 67.3 | 30.1 | 7.5 | 7.5 | 4.2 |
| 20,630 | 141 | 9695.5 | 67.5 | 12.0 | 7.4 | 15.9 | 4.2 |
| 10,920 | 251 | 5099.0 | 189.7 | 41.8 | 7.3 | 7.3 | 5.2 |
| 20,050 | 1 | 6261.7 | 87.5 | 39.3 | 6.9 | 12.6 | 4.5 |
| 30,600 | 2 | 10703.8 | 139.5 | 34.3 | 5.7 | 10.9 | 4.9 |
| 21,730 | 148 | 1871.0 | 28.6 | 24.1 | 5.5 | 11.7 | 3.4 |
| 60,120 | 34 | 2022.0 | 12.0 | 40.1 | 5.0 | 17.9 | 2.5 |
| 60,610 | 23 | 7067.0 | 77.0 | 18.8 | 4.9 | 21.1 | 4.3 |
| 10,320 | 221 | 6037.0 | 93.5 | 14.9 | 4.9 | 9.1 | 4.5 |
| 20,590 | 126 | 546.5 | 4.5 | 35.0 | 4.8 | 9.3 | 1.5 |
| 60,051 | 2 | 425.0 | 1.1 | 42.7 | 4.6 | 10.8 | 0.1 |
| 60,113 | 34 | 2790.0 | 18.4 | 34.0 | 4.6 | 12.6 | 2.9 |
| 30,320 | 228 | 2443.5 | 53.9 | 21.8 | 4.6 | 7.6 | 4.0 |
| 20,820 | 702 | 1377.0 | 10.5 | 26.8 | 4.2 | 7.3 | 2.4 |
| 51,130 | 1 | 8174.0 | 58.9 | 20.6 | 3.9 | 10.7 | 4.1 |
| 40,130 | 111 | 11422.5 | 278.2 | 11.1 | 3.7 | 3.7 | 5.6 |
| 30,021 | 10 | 6327.0 | 127.3 | 14.6 | 3.6 | 9.1 | 4.8 |
| 60,040 | 2 | 595.0 | 3.8 | 46.8 | 3.6 | 11.3 | 1.3 |
| 10,110 | 328 | 2374.5 | 63.2 | 22.0 | 3.6 | 7.2 | 4.1 |
| 60,081 | 34 | 3360.5 | 35.8 | 25.0 | 3.5 | 15.0 | 3.6 |
| 30,001 | 10 | 2779.5 | 29.1 | 26.6 | 3.3 | 11.8 | 3.4 |
| 50,050 | 1 | 1807.5 | 4.4 | 41.5 | 3.3 | 10.5 | 1.5 |
| 40,440 | 308 | 985.0 | 35.3 | 29.8 | 3.2 | 8.1 | 3.6 |
| 20,322 | 125 | 936.0 | 7.8 | 30.2 | 3.1 | 4.2 | 2.1 |
| 30,002 | 10 | 1106.5 | 3.6 | 34.9 | 2.9 | 6.5 | 1.3 |
| 50,160 | 21 | 1384.5 | 7.4 | 38.4 | 2.7 | 14.3 | 2.0 |
| 40,360 | 129 | 14665.0 | 531.8 | 9.8 | 2.7 | 2.7 | 6.3 |

| | | | | | | | |
|---|---|---|---|---|---|---|---|
| 30,290 | 231 | 3746.0 | 50.4 | 18.4 | 2.6 | 6.3 | 3.9 |
| 20,440 | 141 | 5200.0 | 167.1 | 16.0 | 2.4 | 6.7 | 5.1 |
| 20,721 | 4 | 936.0 | 7.4 | 42.4 | 2.4 | 8.7 | 2.0 |
| 70,410 | 240 | 1396.3 | 33.6 | 66.8 | 2.2 | 5.4 | 3.5 |
| 70,020 | 36 | 1288.5 | 8.5 | 39.1 | 2.1 | 10.1 | 2.1 |
| 40,690 | 114 | 1829.0 | 60.5 | 24.3 | 2.0 | 2.0 | 4.1 |
| 50,132 | 21 | 2737.5 | 14.4 | 34.2 | 2.0 | 14.0 | 2.7 |
| 20,210 | 716 | 790.0 | 34.1 | 31.7 | 1.9 | 1.9 | 3.5 |
| 20,840 | 702 | 967.5 | 8.5 | 29.9 | 1.8 | 4.2 | 2.1 |
| 21,071 | 35 | 1955.0 | 15.3 | 29.8 | 1.8 | 7.1 | 2.7 |
| 40,630 | 119 | 2655.5 | 75.4 | 21.1 | 1.7 | 1.7 | 4.3 |
| 20,681 | 141 | 3348.5 | 46.2 | 19.2 | 1.7 | 1.7 | 3.8 |
| 50,270 | 21 | 709.0 | 19.0 | 49.0 | 1.6 | 1.6 | 2.9 |
| 20,610 | 141 | 3030.5 | 18.0 | 20.0 | 1.6 | 4.0 | 2.9 |
| 20,411 | 709 | 2477.5 | 131.2 | 21.7 | 1.6 | 1.6 | 4.9 |
| 70,090 | 32 | 4184.5 | 21.9 | 31.5 | 1.6 | 5.8 | 3.1 |
| 10,930 | 243 | 1523.0 | 19.8 | 25.9 | 1.4 | 6.6 | 3.0 |
| 20,190 | 141 | 5081.0 | 142.3 | 16.1 | 1.3 | 8.8 | 5.0 |
| 20,800 | 148 | 1351.5 | 13.4 | 26.9 | 1.2 | 3.9 | 2.6 |
| 60,982 | 237 | 324.5 | 1.1 | 44.6 | 1.2 | 2.1 | 0.1 |
| 21,060 | 35 | 1007.5 | 5.9 | 41.6 | 1.1 | 5.9 | 1.8 |
| 40,400 | 114 | 1745.0 | 63.0 | 24.7 | 1.1 | 1.1 | 4.1 |
| 20,650 | 35 | 2622.5 | 17.8 | 21.2 | 1.0 | 6.5 | 2.9 |
| 21,351 | 4 | 1157.5 | 6.4 | 28.3 | 1.0 | 2.5 | 1.9 |
| 10,570 | 321 | 1784.0 | 37.9 | 24.5 | 1.0 | 3.8 | 3.6 |
| 10,580 | 242 | 942.5 | 7.5 | 30.2 | 1.0 | 3.1 | 2.0 |
| 20,242 | 135 | 974.5 | 17.4 | 29.9 | 0.8 | 0.8 | 2.9 |
| 21,072 | 35 | 1093.0 | 9.6 | 40.8 | 0.8 | 2.0 | 2.3 |
| 40,100 | 126 | 884.0 | 7.1 | 30.7 | 0.7 | 1.8 | 2.0 |
| 70,361 | 810 | 903.5 | 13.5 | 30.5 | 0.7 | 2.7 | 2.6 |
| 20,830 | 702 | 789.5 | 1.8 | 31.8 | 0.6 | 2.2 | 0.6 |
| 70,040 | 36 | 1561.5 | 4.2 | 25.6 | 0.6 | 2.4 | 1.4 |
| 21,082 | 35 | 756.0 | 3.6 | 44.5 | 0.6 | 1.9 | 1.3 |
| 21,081 | 35 | 962.5 | 3.3 | 42.1 | 0.6 | 2.5 | 1.2 |
| 51,000 | 161 | 1093.0 | 14.3 | 35.0 | 0.6 | 4.1 | 2.7 |
| 30,471 | 10 | 2624.5 | 33.5 | 21.2 | 0.6 | 2.3 | 3.5 |
| 20,421 | 154 | 4421.5 | 141.4 | 24.5 | 0.5 | 4.8 | 5.0 |
| 21,021 | 4 | 607.0 | 4.7 | 50.6 | 0.5 | 2.6 | 1.5 |
| 21,012 | 4 | 1038.0 | 4.3 | 45.0 | 0.4 | 1.5 | 1.5 |
| 30,130 | 232 | 1398.0 | 16.1 | 26.6 | 0.3 | 0.3 | 2.8 |
| 20,060 | 1 | 6463.5 | 23.2 | 19.5 | 0.3 | 0.3 | 3.1 |
| 60,371 | 608 | 537.5 | 5.6 | 35.2 | 0.3 | 0.3 | 1.7 |
| 20,081 | 3 | 6758.0 | 132.5 | 19.1 | 0.3 | 0.3 | 4.9 |
| 20,620 | 141 | 1317.5 | 8.5 | 27.2 | 0.3 | 1.2 | 2.1 |

| | | | | | | | |
|---|---|---|---|---|---|---|---|
| 40,210 | 114 | 1529.0 | 28.5 | 25.8 | 0.3 | 0.3 | 3.4 |
| 50,131 | 21 | 2901.0 | 16.8 | 33.6 | 0.3 | 1.8 | 2.8 |
| 60,560 | 131 | 744.0 | 6.0 | 32.3 | 0.2 | 1.3 | 1.8 |
| 70,010 | 36 | 1341.5 | 10.2 | 33.2 | 0.2 | 2.3 | 2.3 |
| 60,282 | 237 | 634.5 | 4.8 | 33.7 | 0.2 | 2.3 | 1.6 |
| 50,092 | 21 | 3466.0 | 41.8 | 28.8 | 0.2 | 1.8 | 3.7 |
| 10,831 | 244 | 805.5 | 9.3 | 31.6 | 0.2 | 1.7 | 2.2 |
| 60,200 | 1 | 5718.5 | 32.6 | 15.3 | 0.2 | 0.9 | 3.5 |
| 10,030 | 2 | 1017.0 | 5.6 | 29.5 | 0.2 | 0.2 | 1.7 |
| 40,572 | 229 | 1343.5 | 14.7 | 27.0 | 0.2 | 0.2 | 2.7 |
| 50,660 | 158 | 365.0 | 4.0 | 38.5 | 0.2 | 1.6 | 1.4 |
| 70,142 | 32 | 6409.5 | 67.7 | 26.8 | 0.2 | 0.2 | 4.2 |
| 70,030 | 36 | 1439.5 | 15.1 | 38.0 | 0.2 | 1.6 | 2.7 |
| 30,480 | 10 | 1805.5 | 16.9 | 24.4 | 0.1 | 0.1 | 2.8 |
| 50,211 | 151 | 2356.0 | 21.3 | 28.1 | 0.1 | 1.4 | 3.1 |
| 20,600 | 141 | 2237.5 | 9.6 | 28.6 | 0.1 | 0.1 | 2.3 |
| 60,300 | 613 | 1681.5 | 14.1 | 25.0 | 0.1 | 0.1 | 2.6 |
| 70,002 | 36 | 697.5 | 1.8 | 52.5 | 0.1 | 1.2 | 0.6 |
| 60,093 | 34 | 1420.5 | 5.3 | 44.4 | 0.1 | 0.6 | 1.7 |
| 70,211 | 241 | 706.5 | 4.9 | 38.7 | 0.1 | 0.1 | 1.6 |
| 70,160 | 32 | 2796.7 | 18.1 | 41.9 | 0.1 | 0.1 | 2.9 |
| 60,082 | 34 | 2832.5 | 9.2 | 28.7 | 0.1 | 0.1 | 2.2 |
| 70,141 | 32 | 4633.5 | 28.9 | 30.4 | 0.1 | 0.1 | 3.4 |
| 21,024 | 4 | 192.5 | 0.7 | 51.9 | 0.1 | 0.6 | −0.3 |
| 20,722 | 4 | 1113.0 | 4.0 | 40.6 | 0.1 | 0.1 | 1.4 |
| 50,060 | 1 | 1273.5 | 4.5 | 39.2 | 0.1 | 0.1 | 1.5 |
| 10,020 | 2 | 1017.0 | 2.7 | 29.5 | 0.0 | 0.0 | 1.0 |

Linear regression is run on statistical software STATA ®, the command used was *regress Crash_frequency_per_km lnPC_MI_LN_HR SpeedV85*; the results from the calibration produce the following coefficients:

| Crash_frequency | Coef | Std. Error | t | P > \|t\| | 2.5% | 97.5% |
|---|---|---|---|---|---|---|
| ln (PC_MI_LN_HR) | 4.59 | 0.78 | 5.88 | 0.00 | 3.05 | 6.14 |
| Speed V85 | 0.29 | 0.12 | 2.48 | 0.01 | 0.06 | 0.52 |
| Constant | −16.16 | 5.34 | −3.03 | 0.00 | −26.72 | −5.60 |

Therefore, the equations for collision frequency calibrates as:

$$\text{Frequency} = 4.59319^{*} \ \ln\left(\text{PC\_MI\_LN\_HR}\right) + 0.28776^{*} \ \text{Speed\_V85} - 16.16$$

## References


Amador, L., Mohammadi, A., Abu-Samra, S., & Maghsoudi, R. (2020). Resilient storm pipes: a multi-stage decision support system. *Structure and Infrastructure Engineering, 16*(5), 847–859.

Amador-Jimenez, L., & Mrawira, D. (2009). Roads performance modeling & management system from two condition data points: Case study of Costa Rica. *Journal of Transportation Engineering, American Society of Civil Engineering, 135*(2), 999–1007.

Amador-Jimenez, L., & Mrawira, D. (2011a). Capturing variability in pavement performance models from sufficient time-series predictors: A case study of the New Brunswick Road Network. *Canadian Journal of Civil Engineering, 38*(2), 210–220.

Amador-Jimenez, L., & Mrawira, D. (2011b). Reliability-based initial pavement performance deterioration modelling. *International Journal of Pavement Engineering, 12*, 177–186.

Amador-Jimenez, L., & Serrano, L. (2017). Pavement management: A service-based optimal allocation of road's interventions. *International Journal of Sustainable Development and Planning, 12*(6), 1096–1106.

American Association of State Highway and Transportation Officials (AASHTO). (2008). Mechanistic-empirical pavement design guide: A manual of practice.

Ens, A. (2012). Development of a flexible framework for deterioration modelling in infrastructure asset management, University of Toronto, (doctoral dissertation).

Federal Transit Administration (FTA). (2008). Transit State of Good Repair, Beginning the Dialogue, U.S. Department of Transpiration. Retrieved on Jan.22, 2019 from https://www.transit.dot.gov/TAM/TSGRbeginningthedialogue

Gkountis, I. (2014). Infrastructure performance assessment of subway networks. (M.A. Sc. Thesis), Concordia University.

Hall, K.T. & Correa-Muñoz, C.E.(1999). Estimation of present serviceability Infex from international roughness index. Transportation research record. Journal of the Transportation Research Board of the National Academies. Number 1655, pp. 9399.

IBM (2020). IBM-SPSS software, https://www.ibm.com/analytics/spss-statistics-software

Infrastructure Canada. (2016). Average expected useful life of new publicly owned public transit assets. Retrieved on Jan. 22, 2019 from https://www150.statcan.gc.ca.

Mathavan, S. (2014). Use of a self-organizing map for crack detection in highly textured pavement images. *Journal of Infrastructure Systems.*

MathWorks. (2020). MATLAB. https://www.mathworks.com/products/matlab.html

Minitab.(2020). Minitab. https://www.minitab.com/en-us/

Mohammadi, A., Amador-Jimenez, L., & Elsaid, F. (2019). Simplified pavement performance modeling with only two-time series observations: A case study of Montreal Island. *Journal of Transportation Engineering, Part B: Pavements, 145*(4), 05019004.

Mohammadi, A., Amador-Jimenez, L., & Nasiri, F. (2020). Reliable, effective, and sustainable urban railways: A model for optimal planning and asset management. *Journal of Construction Engineering and Management, 146*(6), 04020057.

NCTR. (2016). *State of good repair performance measures: assessing asset condition, age, and performance data* (No. CUTR-NCTR-RR-2016-09).

Ontario Ministry of Transportation. (2013), Pavement design and rehabilitation manual, second edition, www.bv.transports.gouv.qc.ca/mono/1165561.pdf, (1 October 2018).

Paterson, W. D. O., & Attoh-Okine, B. (1992). Summary of Models of Pavement Road, Deterioration Based on HDM-III. Transportation Research Record: Journal of the Transportation Research Board, 1344, 99–105.

RStudio.(2020). https://rstudio.com/

RTA (Regional Transportation Authority). (2014). Capital condition assessment report, Department of Finance & Performance Management, Chicago, December 2014. Retrieved on Jan. 22, 2019 from https://www.rtachicago.org

Semaan, N. (2011). Structural performance model for subway networks. Ph.D. thesis. Department of Building, Civil, and Environmental Engineering, Concordia University.

Sobanjo, J. O. (2011). State transition probabilities in bridge deterioration based on Weibull sojourn times. *Structure and Infrastructure Engineering, 7*(10), 747–764.
Thomas, O., & Sobanjo, J. (2016). Semi-Markov models for the deterioration of bridge elements. *Journal of Infrastructure Systems, 22*(3), 04016010.
Toronto Hydro Electric System. (2009). Useful life of assets, Kinectrics Inc. Report No: K-418021-RA-0001-R002. Retrieved on Jan. 22, 2019 from https://www.torontohydro.com.
Zareie, A., Amin, M. S. R., & Amador-Jiménez, L. E. (2016). Thornthwaite moisture index modeling to estimate the implication of climate change on pavement deterioration. *Journal of Transportation Engineering, 142*(4), 04016007.

# Chapter 4
# Infrastructure Interventions

## 4.1 What Is the Effectiveness of the Interventions?

One of the most important aspects of the IAM system is the capability of assessing the effectiveness of the interventions in order to select the most cost-effective alternatives. Effectiveness is expected in the form of any of the following effects:

- Extension on the lifespan of the asset presents as age (i.e., year)
- Rejuvenation on some indicator characteristic of its functional or structural capability (e.g., increasing PCI or dropping IRI for pavement)
- Reduction on the rate of deterioration

Intervention effectiveness is typically visualized on a trendline graph by observing improvement in the condition indicator on the *y*-axis, and extension in the asset life span (i.e., growing remaining service life [RSL]) on the *x*-axis (Fig. 4.1). The effectiveness of an intervention depends on the type and quality of the implementation method, timing, materials, equipment/tools, and workers' skills. Therefore, it varies location by location, and for this reason, asset owners are recommended to estimate the effectiveness of interventions locally. The gain of each action can be impacted by the time of implementation because of the condition state that the asset is currently at. For instance, preventive maintenance is more effective when the asset is still relatively new. Hence, the specific condition range (or age range) where each intervention is more effective must also be identified.

Estimation of the effectiveness for any given intervention starts by looking into the historical trend of decay of a given indicator before, during, and after the moment of time the intervention is deployed (Fig. 4.1).



A. Mohammadi, L. Amador Jimenez, *Asset Management Decision-Making For Infrastructure Systems*, https://doi.org/10.1007/978-3-030-97614-9_4

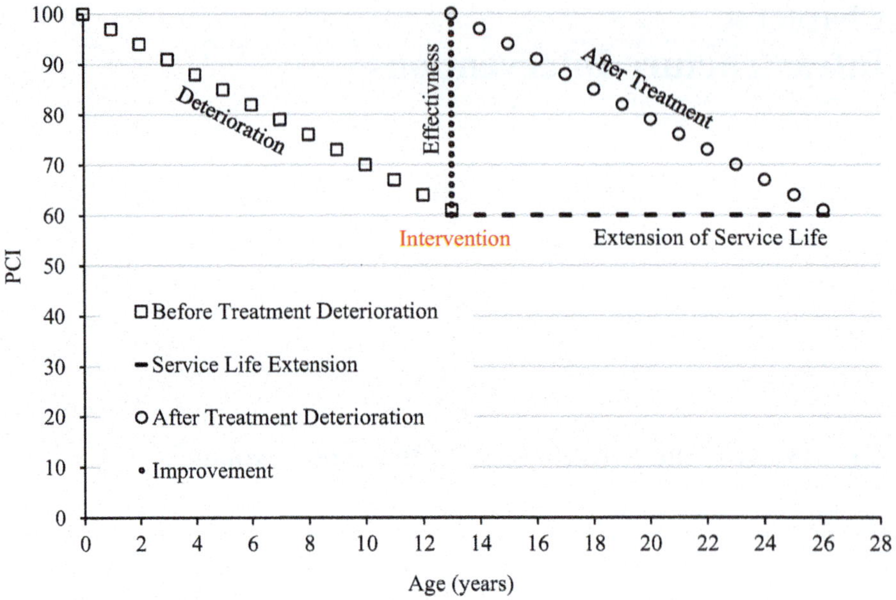


**Fig. 4.1** Intervention effectiveness

## 4.2 Estimating Effectiveness

### *4.2.1 Deterministic Methods*

The first step when measuring effectiveness using a deterministic method is to produce the best fit curve for the before-and-after treatment conditions. A simplifying assumption is that assets will follow the same trend of deterioration after a given intervention (Fig. 4.1), which may or may not happen. Let's assume an agency did minor rehabilitation and PCI improved from 55 to 85. Figure 4.2 presents the deterioration model for this road. Effectiveness for this maintenance intervention (i.e., minor rehabilitation) can be estimated from two perspectives of condition improvement (i.e., PCI) and life span extension (i.e., age).

An increase of PCI from 55 to 85 means a benefit of 30 PCI. To estimate life span extension, the asset rejuvenation related to the improvement on PCI value (before and after treatment) should be estimated using the deterioration curve (given equation), and for this case would be (3.5–1) or 2.5 years. It means that the asset has rejuvenated 2.5 years, and its apparent age is no longer 3.5 years but rather 1 year. We call this age "apparent" because it is not the real age from the time of construction, but rather the corresponding value to the current level of condition.

**Example 4.1**
A municipality recently tested two types of intervention for the pavement to collect data for developing an AM platform. Table 4.1 presents before and after treatment inspection (PCI) results. What is the benefit (per year) of these interventions? The deterioration model for this city is given in Fig. 4.2.

***Solution*** In this example, for each intervention more than one sample was collected and therefore an average can be used. Using the deterioration model, the corresponding age can be calculated mathematically, and an average of 1.05 is assigned for preventive maintenance, which typically shows that this intervention is either non-efficient or the PCI indicator should be replaced with more detailed indicators that match the damage observed and intervention applied. For reconstruction, all samples reflected after intervention PCI value of 100 (Table 4.2), which is

**Table 4.1** Example 4.1 dataset

| Treatment type | Implementation range (PCI) | PCI | |
|---|---|---|---|
| | | Before treatment | After treatment |
| Preventive maintenance | 70–90 | 83 | 98 |
| | | 78 | 95 |
| | | 72 | 88 |
| Reconstruction | 0–30 | 27 | 100 |
| | | 23 | 100 |
| | | 21 | 100 |

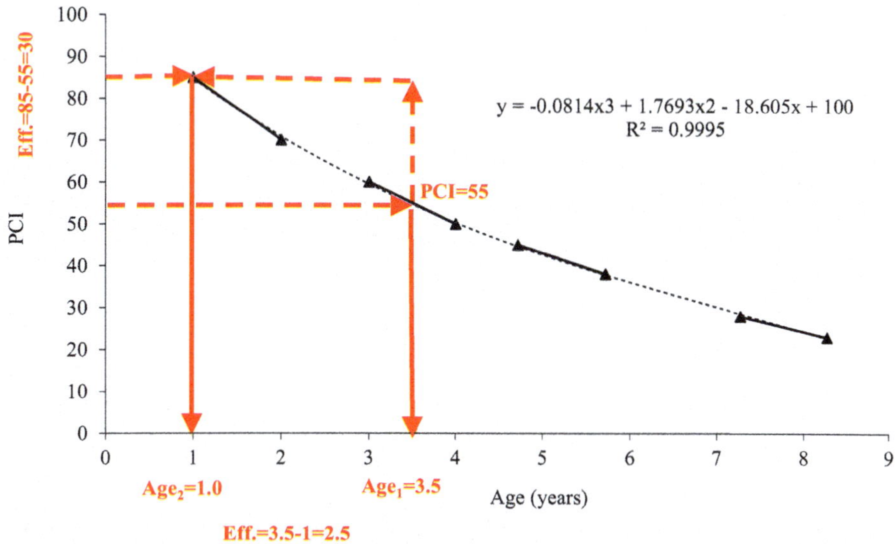


**Fig. 4.2** Example for estimating intervention effectiveness for a road pavement

**Table 4.2** Example 4.1 solution

| | | PCI | | Corresponding age | | Effectiveness | Effectiveness |
|---|---|---|---|---|---|---|---|
| Treatment type | Implementation range | Before treatment | After treatment | Before treatment | After treatment | Age | Age |
| Preventive maintenance | 70–90 | 83 | 98 | 1.11 | 0.20 | 0.91 | 1.05 |
| | | 78 | 95 | 1.46 | 0.37 | 1.09 | |
| | | 72 | 88 | 1.95 | 0.79 | 1.16 | |
| Reconstruction | 0–30 | 27 | 100 | | | As new | As new |
| | | 23 | 100 | | | As new | |
| | | 21 | 100 | | | As new | |

expected, and so the effectiveness is a full reset to brand new asset. Mathematically, we may estimate a gain per age for each sample after applying reconstruction; however, from an IAM perspective, it has no meaning as the asset will be converted to a brand-new asset.

The reader must consider that it is assumed that the second condition assessment (i.e., after treatment) is done in a short span apart from the previous assessment, and so the assessment on the year of deployment of the intervention carries a small (and for the purpose of this book, negligible) decay in condition. Otherwise, this may be considered in estimating effectiveness. For instance, if there is a one-year interval between two observations, we may add one year to this estimate as assets are continuously deteriorated before or after treatment.

Capturing the before and after shift on the deterioration trend of a given indicator sometimes is not sufficient, because the effectiveness of treatment also may involve a slow down on the rate (slope) of deterioration. For this reason, it is ideal to count on multiple observations after the deployment of the intervention.

One final consideration, as we are forecasting future levels of deterioration, there is an increasing degree of uncertainty associated with the decay of a given condition indicator (structural or functional) for both before and after any intervention (Fig. 4.3).

**Example 4.2**

Consider the following dataset for the observed degree of corrosion on the superstructure of a steel bridge: from 1991 through 1999 no intervention was practiced, and the bridge's superstructure corrosion increased. In the year 1999, the budget was allocated for the removal of corrosion (replacement of steel members) and anti-corrosion painting of the bridge during the summer of the year 2000. From the year 2000 onwards, the "corrosion defects" were removed (hence corrosion area drops to zero) but new corrosion began due to the environmental effects of the region (Table 4.3).

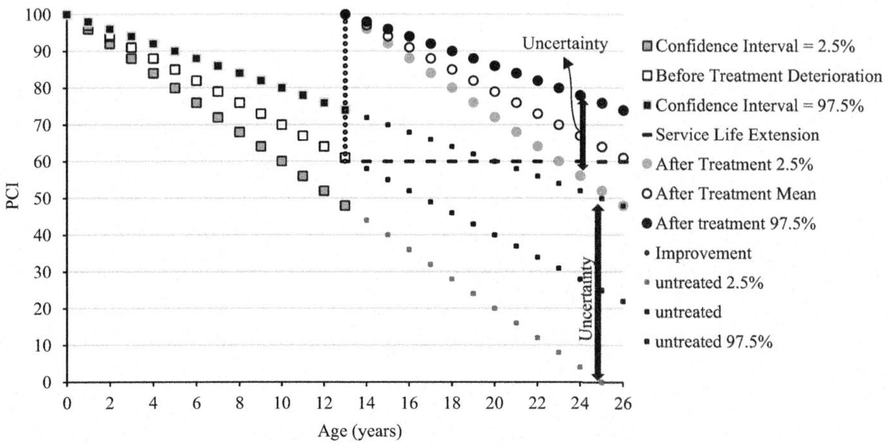


**Fig. 4.3** Uncertainty before and after an intervention

**Table 4.3** Corrosion for a bridge: Historical progression, forecast and observed after treatment trend

| Phenomena | Year | Corrosion (% of the surface) |
|---|---|---|
| Corrosion progression | 1991 | 1.00 |
| | 1992 | 2.25 |
| | 1993 | 5.51 |
| | 1994 | 7.75 |
| | 1995 | 9.88 |
| | 1996 | 12.01 |
| | 1997 | 15.03 |
| | 1998 | 18.02 |
| | 1999 | 21.07 |
| Projected corrosion if no intervention is done | 2000 | 22.60 |
| | 2001 | 25.07 |
| | 2002 | 27.54 |
| | 2003 | 30.01 |
| | 2004 | 32.48 |
| | 2005 | 34.95 |
| Observed corrosion after intervention (removal and paint) | 2000 | 0.00 |
| | 2001 | 0.50 |
| | 2002 | 1.10 |
| | 2003 | 1.60 |
| | 2004 | 2.21 |
| | 2005 | 2.73 |

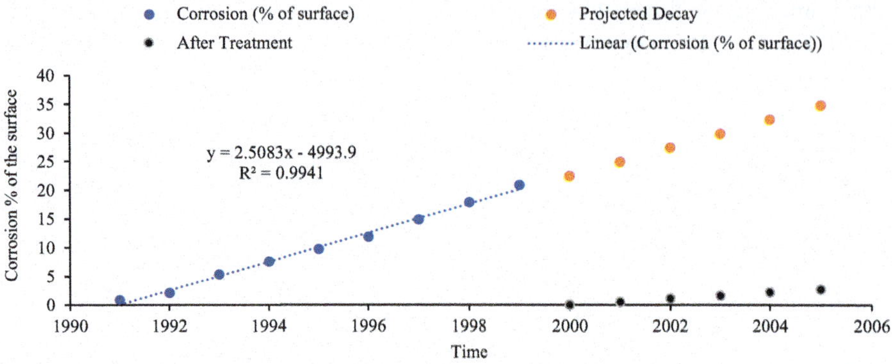


**Fig. 4.4** Before and after trends for one bridge

By looking at the information, one can produce a simple linear regression model (Fig. 4.4) and estimate the best fit; let's assume that the equation is given:

$$\text{Corrosion } (\%) = 2.5083^{*}\ (\text{year}) - 4993.9$$

This simple equation can be used to quickly forecast the expected degree of corrosion (% of the area) if no intervention had occurred. However, there was an intervention that resettled the corrosion levels to zero and slowed down the rate of corrosion because of the anticorrosion painting applied.

Figure 4.4 shows that the corrosion dropped to zero from 21.07%; however, to estimate lifespan expansion, a new deterioration trend should be considered. If the original decay line is followed, the lifespan extension would be almost 9 years, while the new trend shows the bridge will reach the almost same corrosion level after 30 years using this updated equation:

$$\text{Corrosion } (\%) = 0.5509^{*} \text{ (year)} - 1101.7$$

### 4.2.2 Stochastic Methods

The idea of developing TPM can be also applied for effectiveness where the matrix probabilities indicate the chance of improving asset condition from one state to the upper (i.e., better state). Thus, the same methodology can be used to develop a matrix, and by comparing and counting samples before and after treatments, the left side of the matrix will be extracted. This matrix will be able to provide the probability of improving (i.e., effectiveness) asset state by applying a maintenance intervention, however, cannot directly give an idea about lifespan expansion. To cover this limitation, a customized TPM is developed as below:

Let us use the source dataset from which Fig. 3.15 average points were estimated (Fig. 4.5).

In this customized matrix, the state ranges are defined using bridge apparent ages. The model equation is used to obtain matrix ranges: to do so, simply replace integer time values (i.e., 1,2, 3,. . .9) called apparent age into the equation to obtain the higher end of the corrosion values (Table 4.4).

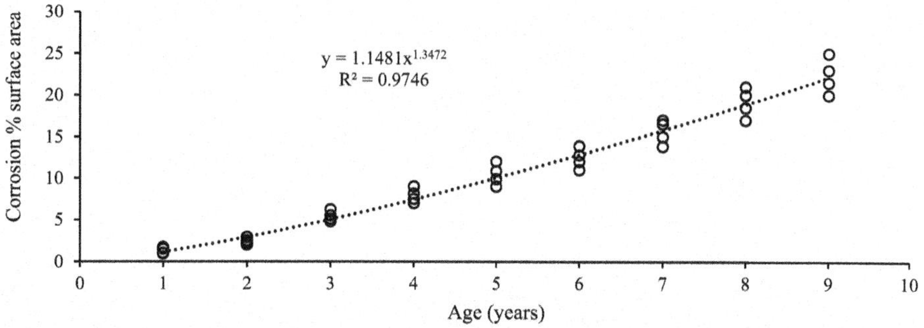


**Fig. 4.5** Deterioration model developed from historical observations of corrosion progression

**Table 4.4** Definition of lower and higher boundary values for corresponding apparent ages

| Time | Corrosion | Lower end | Higher end | Apparent age |
|---|---|---|---|---|
| 1 | 1.15 | 0 | 1.15 | 1 |
| 2 | 2.92 | 1.16 | 2.92 | 2 |
| 3 | 5.04 | 2.93 | 5.04 | 3 |
| 4 | 7.43 | 5.05 | 7.43 | 4 |
| 5 | 10.04 | 7.44 | 10.04 | 5 |
| 6 | 12.83 | 10.05 | 12.83 | 6 |
| 7 | 15.79 | 12.84 | 15.79 | 7 |
| 8 | 18.91 | 15.80 | 18.91 | 8 |
| 9 | 22.16 | 18.92 | 22.16 | 9 |

**Table 4.5** Empty transition probability matrix: Higher boundary value from deterministic curve

| From→ To↓ | Age | 1 | 2 | 3 | 4 | 5 | 6 | 7 | 8 | 9 |
|---|---|---|---|---|---|---|---|---|---|---|
| Age | Corrosion | 0–1.15 | 1.16–2.92 | 2.93–5.04 | 5.05–7.43 | 7.44–10.04 | 10.05–12.83 | 12.84–15.79 | 15.80–18.91 | 18.92–22.16 |
| 1 | 0–1.15 | | | | | | | | | |
| 2 | 1.16–2.92 | | | | | | | | | |
| 3 | 2.93–5.04 | | | | | | | | | |
| 4 | 5.05–7.43 | | | | | | | | | |
| 5 | 7.44–10.04 | | | | | | | | | |
| 6 | 10.05–12.83 | | | | | | | | | |
| 7 | 12.84–15.79 | | | | | | | | | |
| 8 | 15.80–18.91 | | | | | | | | | |
| 9 | 18.92–22.16 | | | | | | | | | |

Using the previous values, establish the probabilistic matrix as illustrated in Table 4.5. Notice how the main diagonal is shaded in gray to denote the fact that any value on the main diagonal reflects a segment that has neither deteriorated nor improved. Although we could use the main diagonal to capture deterioration trends by doing the procedure that follows, we will only concentrate on capturing improvement trends from the data we have observed. To do so, populate the probabilistic matrix by observing the changes in bridges that had actually received an intervention. The asset owner is expected to have a historical record (since the action was applied recently) of which intervention was applied in which level of corrosion. Then, this matrix will help to identify the right action as well as the right time to implement each.

**Table 4.6** Before and after corrosion % on 10 bridges and corresponding rejuvenation age

| | Before | After | Back to age |
|---|---|---|---|
| Time | 19.97 | 1.75 | 2 |
| Bridge 1 | 19.08 | 1.55 | 2 |
| Bridge 2 | 21.34 | 2.5 | 2 |
| Bridge 3 | 22.09 | 2.4 | 2 |
| Bridge 4 | 22.04 | 3.5 | 3 |
| Bridge 5 | 19.4 | 4.05 | 3 |
| Bridge 6 | 19.88 | 5.4 | 4 |
| Bridge 7 | 18.99 | 1.25 | 2 |
| Bridge 8 | 21.25 | 0.8 | 1 |
| Bridge 9 | 22 | 4.5 | 3 |
| Bridge 10 | 21.95 | 0.9 | 1 |

**Table 4.7** Transition probability matrix: 10 bridges, rejuvenation

| From→ To↓ | Age | 1 | 2 | 3 | 4 | 5 | 6 | 7 | 8 | 9 |
|---|---|---|---|---|---|---|---|---|---|---|
| Age | Corrosion | 0–1.15 | 1.16–2.92 | 2.93–5.04 | 5.05–7.43 | 7.44–10.04 | 10.05–12.83 | 12.84–15.79 | 15.80–18.91 | 18.92–22.16 |
| 1 | 0–1.15 | | | | | | | | | |
| 2 | 1.16–2.92 | | | | | | | | | |
| 3 | 2.93–5.04 | | | | | | | | | |
| 4 | 5.05–7.43 | | | | | | | | | |
| 5 | 7.44–10.04 | | | | | | | | | |
| 6 | 10.05–12.83 | | | | | | | | | |
| 7 | 12.84–15.79 | | | | | | | | | |
| 8 | 15.80–18.91 | | | | | | | | | |
| 9 | 18.92–22.16 | 2 | 5 | 3 | 1 | | | | | |

For simplicity, we continue using the corrosion % area as an indicator of the condition for the steel bridge's superstructure. Table 4.6 provides us with the before and after values observed for the 10 bridges from which we had observations.

When we bring the count of how many bridges had rejuvenated into the probability matrix, all counts from Table 4.6 will be located at the departure apparent age of 9 (the before intervention value) in Table 4.7. The treatment effectiveness is captured by the distance from the main diagonal to the corresponding after-treatment cell.

This effectiveness produces a rejuvenation with probabilities of 10% (1 out of 10), 30%, 50%, and 20% for resetting the curve back to ages 4,3,2 and 1, respectively, as shown in Table 4.8. The rate of deterioration will continue following the previously estimated trend:

**Table 4.8** After-treatment deterioration trends: Four probabilistic outcomes

| Year | | Age 4 | | Age 3 | | Age 2 | | Age 1 |
|---|---|---|---|---|---|---|---|---|
| (Time) | Rejuvenate | 10.0% | Rejuvenate | 30.0% | Rejuvenate | 50.0% | Rejuvenate | 20.0% |
| 9 | 4 | 6.24 | 3 | 5.20 | 2 | 3.01 | 1 | 1.18 |
| 10 | 5 | 10.35 | 4 | 7.66 | 3 | 5.20 | 2 | 3.01 |
| 11 | 6 | 13.23 | 5 | 10.35 | 4 | 7.66 | 3 | 5.20 |
| 12 | 7 | 16.29 | 6 | 13.23 | 5 | 10.35 | 4 | 7.66 |
| 13 | 8 | 19.50 | 7 | 16.29 | 6 | 13.23 | 5 | 10.35 |

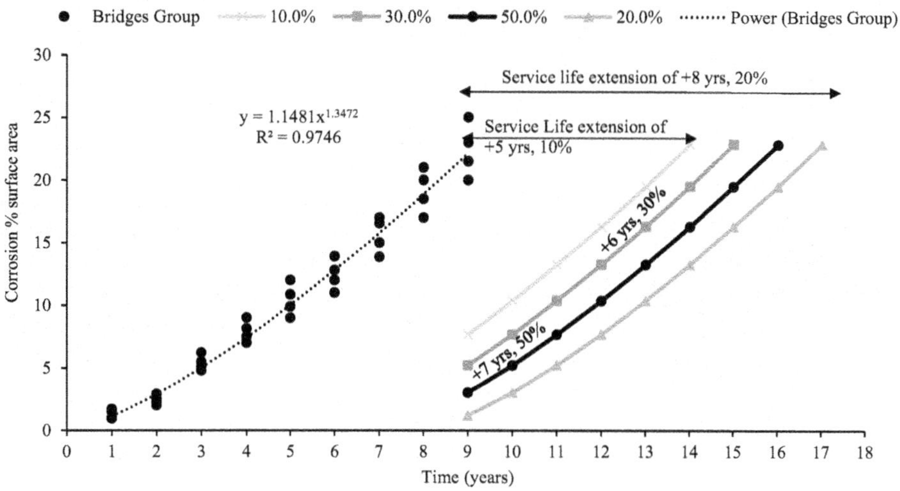


**Fig. 4.6** After-treatment probabilistic performance

**Table 4.9** Weighted average of life extension

| Probability | Life extension | Probability*Life extension |
|---|---|---|
| 10 | 5 | 0.5 |
| 30 | 6 | 1.8 |
| 50 | 7 | 3.5 |
| 20 | 8 | 1.6 |
| Weighted average= | | **7.4** |

$$\%\text{corrosion} = 1.1481\,(\text{apparent age})^{1.3472}$$

The previous trends can be better visualized (Fig. 4.6) on a graph with the original trend, the rejuvenation reset, and the after-treatment decay for the four probabilistic outcomes.

In simple words, there is 50% of the effectiveness of the intervention results in a reset to age 2, meaning a 7-year extension of service life. Likewise, there is a 30% chance the rejuvenation results in a 6-year extension of lifespan, a 20% chance for +8 years, and a 10% chance for 5 years extension. The weighted average would result in data shown in Table 4.9. Thus, on average, the observed treatment extends the lifespan of the steel superstructure of the bridges in 7.4 years for this fictional example.

Entire datasets are used on real-life applications to fill out the entire transition probability matrix. The next section presents a complete example of real values collected in New Brunswick roads in Canada. A second case study has been provided in Appendix 3.

## 4.3 Case Study 1: Segments from Routes 1 and 8 of New Brunswick

A database with the surface condition (IRI) and pavement structural capacity (deflection area basin parameter) previously assembled (Amador Jimenez & Mrawira, 2011) were used in this paper. The data contained 6 years of observations in the form of the international roughness index (IRI) and area deflection basin parameter (AREA). A subset of data points was selected from segments receiving surface treatments from 1991 to 1996. This resulted in 45 segments of 161 m (1/10 of a mile) length from two sites: *Group 1*: 3540 meters of control section 18 of route 8, and *Group 2*: 4184 meters from control section 8 of route 1, both receiving surface treatments during 1991 or 1995, correspondingly. All sections came from asphalt-paved arterial roads. However, group 1 had a moisture index of 60 and group 2 of 80, annual freeze-thaw day cycles were similar: 16.81 and 14.29 days per year, correspondingly.

The following steps were used for capturing treatment effectiveness:

1. Gather performance data.
   (a) At least one indicator of surface condition and one of structural capacity
   (b) Obtain a record of treatments
2. Get subsets.
   (a) Pavements treated by each specific treatment type
   (b) With records of performance before and after treatment
3. Divide segments into families of pavements.
4. Develop deterioration curves per pavement family.
   (a) Use time-series analysis, if sufficient data
   (b) Use the apparent age approach, if insufficient time-series data
5. Generate a TPM per group and condition indicator.
   (a) Synchronize each cell movement to coincide with 1 year of deterioration
      (i) Use deterioration curves developed in Step 4
6. Measure treatment effectiveness using the TPM.
   (a) Service life extension (reset value)
   (b) Changes in slopes
   (c) After treatment performance

Ideally, the indicator of surface condition (Step 1a) must be the one used to trigger specific treatments; i.e., cracking for crack sealing, rutting, and cracking for micro-surfacing, etc. New Brunswick Department of Transportation (NBDOT) has a protocol establishing such treatment criteria and operational windows. The TPM produced by the previous procedure is expected to be used in combination with decision rule sets to allocate maintenance and rehabilitation treatments for pavement management.

The following table shows the observed condition and traffic loading per segment; as seen, there was missing data from the falling weight deflectometer (FWD) readings not collected for some of the segments.

Two indicators were used to assess the structural and functional decay of roads: the deflection basin parameter (DBP) (structural strength) and the IRI (road surface smoothness). They are briefly explained in the following sections.

### 4.3.1 *Deflection Basin Parameter (DBP)—Area*

FWD is often used to evaluate the physical properties of pavements. FWD data is primarily used to estimate pavement structural capacity. FWDs impose a load pulse to a pavement surface by dropping a large weight, simulating the load of a vehicle's wheel. Deflection sensors mounted at fixed offsets from the center of a load plate measure the deformation of the pavement in response to the load. Direct analysis of deflection data can be done by using a DBP as a surrogate of pavement strength (Xu et al., 2003). The Area deflection basin parameter (Area) has been used as a proxy of pavement structural capacity in the absence of structural numbers due to the lack of information on the thickness of pavement layers (Amador Jimenez & Mrawira, 2011). Equation 4.1 shows the Area DBP used in this paper.

$$\text{Area} = 6\left[1 + 2\left(\frac{D_1}{D_0}\right) + 2\left(\frac{D_2}{D_0}\right) + \left(\frac{D_3}{D_0}\right)\right] \tag{4.1}$$

Where $D_0$, $D_1$, $D_2$, and $D_3$ are FWD deflection readings at zero offset, first offset, second offset, and third offset geophones, respectively. Theoretically, Area can fluctuate from 36 (strong) to 11.1 (weak); however, observed values for the dataset of this paper ranged from 17.7 to 32.6.

### 4.3.2 *Modeling Deterioration of Area Deflection Basin Parameter: An Apparent Age Approach*

Performance models were developed from cross-sectional data of thousands of segments of roads and longitudinal data for 6 years (1991 to 1996). An apparent age (a surrogate of condition) was correlated with pavement condition (i.e., DBP). Road segments were divided into groups based on traffic intensity or environmental exposure. Further, four qualitative subgroups of the condition were defined (i.e., good, fair, poor, very poor). Finally, segments that showed signs of deterioration for two consecutive years were rearranged in pairs (*initial, final*), and their average condition was calculated.

The procedure started by assuming an apparent age ($AGE_1$) of zero for as-built FWD of 31 (highest observed value); this is called breakpoint one ($BP_1$). This arbitrary assumption was based on the highest observed Area deflection basin value for the network and can be customarily adjusted for other calibrations. The first apparent age ($AGE_1$) to be determined was for the pair of average Area basin points for roads in "Good" condition ($\mu_{\text{initial}}^{\text{Good}}, \mu_{\text{final}}^{\text{Good}}$). This was determined by finding the age value of the second breakpoint ($AGE_2$) that achieved the objective of separating the first pair of average IRI points ($\mu_{\text{initial}}^{\text{Good}}, \mu_{\text{final}}^{\text{Good}}$) by a distance of 1 year, which is the time elapsed between successive condition surveys. The apparent age ($AGE_3$) for the third breakpoint ($BP_3$) used the just established apparent age of the second breakpoint ($AGE_2$) to find the value of the corresponding age of the third breakpoint ($AGE_3$) that achieves a distance of 1 year between the second pair of average fair IRI points ($\mu_{2004}^{\text{Fair}}, \mu_{2006}^{\text{Fair}}$). This procedure continues in this fashion using the average values of the initial and final Area basin for poor and very poor pairs of average pavement conditions until all apparent ages have been established. Equation 4.2 was used to find the apparent age of each breakpoint.

$$\frac{\text{BP}_n - \mu_{\text{final}}}{\left(\frac{\text{BP}_n - \text{BP}_{n+1}}{\text{AGE}_{n+1} - \text{AGE}_n}\right)} - \frac{\text{BP}_n - \mu_{\text{initial}}}{\left(\frac{\text{BP}_n - \text{BP}_{n+1}}{\text{AGE}_{n+1} - \text{AGE}_n}\right)} = 1 \tag{4.2}$$

Where $\text{BP}_n$ represents the breakpoint (i.e., Area basin) corresponding to apparent age $n$; $\text{AGE}_n$ = apparent age $n$; and $\mu_{\text{initial}}$ or $\mu_{\text{final}}$ = the mean condition (Area basin) of the group at any initial or final year. Apparent ages for the breakpoints of the traffic intensity groups were used as a basis to assign apparent ages for different groups. A performance model was built by plotting pairs of apparent ages and breakpoints.

### *4.3.3 Exploratory Analysis*

Box plots of IRI for before, after, and when the treatment was applied, resulting in a clear indication that for group 2 (treated in 1995) the variability of observed roughness diminished after receiving a surface treatment, even though IRI increased from the year before (Fig. 4.7). Before treatment data for group 1 (receiving treatment during 1991) was missing, only after treatment could be observed, noticing a negligible increment on IRI (Fig. 4.4). This drastic difference can be explained by the fact that traffic loading (ESALs) on route 8 control section 18 (group 1) dropped from the year of application (of surface treatment) to the following year from 404,272 to 204,034 ESALs (Table 4.10).

A line plot of individual values of before and after roughness (IRI) confirms that for group 2, the overall variability dropped (comparing before and applied lines), but in addition, those segments with poorer conditions benefited the most (Fig. 4.8). However, such rejuvenation seems to last somewhere about 2 years, by simply

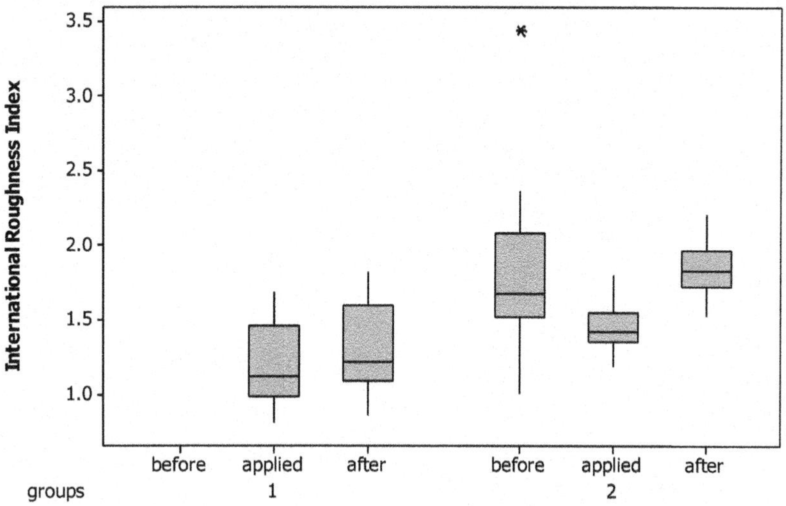


**Fig. 4.7** Box plot of before, after, and at the year of treatment application, groups 1 and 2

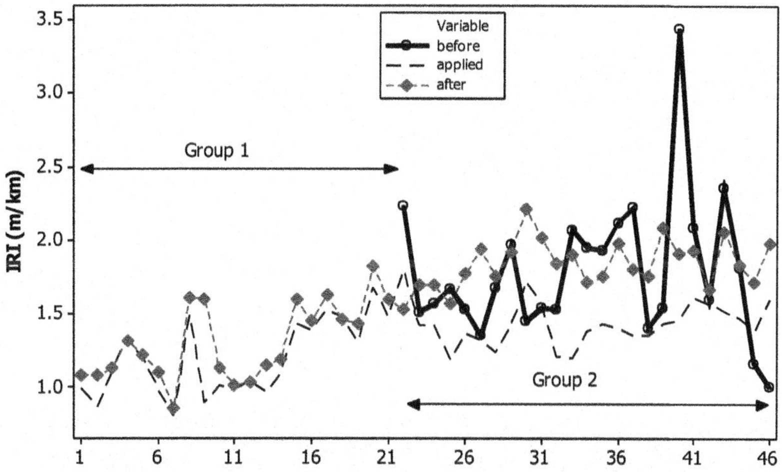


**Fig. 4.8** Profile of surface condition (IRI) before and after a surface treatment

contrasting the line trends of the year of application with the one before and after. For group 1, a very close trend can be observed between the year of application and the one immediately following. Even though the after-line shows some signs of wearing, it is by far much less than that showed by group 2, because of the drop (−49.6%) in traffic loading on group 1 as compared to a 3.86% increase for group 2.

A box plot of the Area basin parameter for group 2 showed that pavement structural deterioration slowed down after receiving a surface treatment (Fig. 4.9), possibly because of waterproofing from sealing surface cracks.

**Table 4.10** Dataset from New Brunswick used in this case study

| | IRI | | | | | | Area Basin–FWD | | | | | | ESALs | |
|---|---|---|---|---|---|---|---|---|---|---|---|---|---|---|
| Year | 1991 | 1992 | 1993 | 1994 | 1995 | 1996 | 1991 | 1992 | 1993 | 1994 | 1995 | 1996 | Before | After |
| 1991 | 2.77 | 1.97 | 2.13 | 2.23 | 1.80 | 1.53 | | | | | | | 404,732 | 204,034 |
| 1991 | 1.84 | 1.36 | 1.61 | 1.51 | 1.42 | 1.70 | 22.03 | 21.69 | 21.18 | 26.59 | 23.86 | 21.82 | 404,732 | 204,034 |
| 1991 | 1.77 | 1.43 | 1.59 | 1.57 | 1.42 | 1.70 | 24.16 | 24.59 | 23.91 | 28.84 | 25.46 | 25.07 | 404,732 | 204,034 |
| 1991 | 1.44 | 1.49 | 1.50 | 1.67 | 1.19 | 1.57 | | | | | | | 404,732 | 204,034 |
| 1991 | 1.52 | 1.41 | 1.50 | 1.53 | 1.37 | 1.77 | 22.63 | 22.74 | 22.44 | 25.08 | 23.22 | 22.13 | 404,732 | 204,034 |
| 1991 | 1.38 | 1.28 | 1.32 | 1.35 | 1.32 | 1.94 | 19.96 | 19.81 | 18.27 | 19.66 | 19.60 | 19.38 | 404,732 | 204,034 |
| 1991 | 1.75 | 1.70 | 1.77 | 1.68 | 1.24 | 1.75 | 22.65 | 22.71 | 21.72 | 23.14 | 20.35 | 20.09 | 404,732 | 204,034 |
| 1991 | 1.98 | 1.88 | 2.01 | 1.97 | 1.43 | 1.92 | 19.43 | 18.36 | 17.68 | 29.07 | 19.50 | 19.29 | 404,732 | 204,034 |
| 1991 | 1.58 | 1.49 | 1.56 | 1.45 | 1.72 | 2.21 | | | | | | | 404,732 | 204,034 |
| 1991 | 1.54 | 1.53 | 1.55 | 1.54 | 1.57 | 2.02 | 22.83 | 23.74 | 21.42 | 24.57 | 23.10 | 21.83 | 404,732 | 204,034 |
| 1991 | 1.63 | 1.57 | 1.55 | 1.53 | 1.21 | 1.84 | 20.48 | 18.69 | 18.61 | 23.17 | 20.26 | 19.65 | 404,732 | 204,034 |
| 1991 | 2.02 | 1.99 | 2.11 | 2.07 | 1.20 | 1.90 | 25.61 | 25.31 | 23.52 | 31.70 | 25.47 | 24.00 | 404,732 | 204,034 |
| 1991 | 1.91 | 1.82 | 1.98 | 1.95 | 1.38 | 1.72 | 23.08 | 23.82 | 21.68 | 25.28 | 22.88 | 21.93 | 404,732 | 204,034 |
| 1991 | 1.75 | 1.57 | 1.66 | 1.93 | 1.43 | 1.75 | | | | | | | 404,732 | 204,034 |
| 1991 | 2.05 | 1.90 | 2.05 | 2.12 | 1.41 | 1.98 | 22.35 | 22.10 | 20.81 | 25.17 | 23.36 | 22.24 | 404,732 | 204,034 |
| 1991 | 2.27 | 2.13 | 2.23 | 2.22 | 1.35 | 1.80 | 23.12 | 25.87 | 22.84 | 25.23 | 24.66 | 23.81 | 404,732 | 204,034 |
| 1991 | 1.47 | 1.39 | 1.38 | 1.40 | 1.35 | 1.75 | 19.21 | 20.44 | 18.31 | 28.18 | 20.11 | 20.00 | 404,732 | 204,034 |
| 1991 | 1.53 | 1.52 | 1.52 | 1.54 | 1.43 | 2.09 | 20.93 | 21.41 | 19.76 | 18.45 | 18.77 | 18.18 | 404,732 | 204,034 |
| 1991 | 2.76 | 2.84 | 3.09 | 3.44 | 1.46 | 1.91 | | | | | | | 404,732 | 204,034 |
| 1991 | 2.06 | 2.15 | 2.26 | 2.09 | 1.61 | 1.93 | 22.69 | 19.86 | 19.62 | 21.54 | 21.94 | 20.94 | 404,732 | 204,034 |
| 1991 | 1.61 | 1.44 | 1.92 | 1.60 | 1.57 | 1.67 | 26.55 | 29.56 | 25.37 | 24.62 | 19.31 | 18.41 | 404,732 | 204,034 |
| 1995 | 2.32 | 2.34 | 2.32 | 2.36 | 1.52 | 2.06 | 21.83 | 22.42 | 21.20 | 23.33 | 21.24 | 19.92 | 556,877 | 578,372 |
| 1995 | 1.83 | 2.24 | 1.86 | 1.81 | 1.47 | 1.83 | 22.95 | 23.57 | 22.20 | 22.77 | 21.75 | 21.09 | 556,877 | 578,372 |
| 1995 | 1.21 | 1.15 | 1.17 | 1.16 | 1.39 | 1.72 | 20.48 | 19.25 | 19.07 | 22.62 | 22.63 | 20.55 | 556,877 | 578,372 |
| 1995 | 1.13 | 1.03 | 1.01 | 1.00 | 1.60 | 1.98 | 23.19 | 24.55 | 23.06 | 26.95 | 20.01 | 18.91 | 556,877 | 578,372 |

| | | | | | | | | | | | | | | |
|---|---|---|---|---|---|---|---|---|---|---|---|---|---|---|
| 1995 | 0.99 | 1.08 | 1.15 | 1.04 | 1.22 | 1.34 | 22.78 | 21.78 | 25.00 | 24.06 | 23.16 | 24.30 | 556,877 | 578,372 |
| 1995 | 0.87 | 1.08 | 0.97 | 0.89 | 1.09 | 1.25 | 23.36 | 22.27 | 30.46 | 25.10 | 23.72 | 24.43 | 556,877 | 578,372 |
| 1995 | 1.12 | 1.13 | 1.06 | 1.05 | 1.14 | 1.44 | 25.58 | 24.73 | 25.70 | 24.23 | 23.06 | 23.64 | 556,877 | 578,372 |
| 1995 | 1.34 | 1.32 | 1.35 | 1.49 | 1.62 | 1.85 | 22.85 | 22.48 | 30.82 | 24.10 | 23.27 | 24.46 | 556,877 | 578,372 |
| 1995 | 1.20 | 1.22 | 1.43 | 1.39 | 1.48 | 1.97 | | | | | | | 556,877 | 578,372 |
| 1995 | 0.98 | 1.10 | 1.08 | 1.14 | 1.24 | 1.86 | 24.88 | 24.30 | 32.63 | 26.97 | 24.63 | 25.03 | 556,877 | 578,372 |
| 1995 | 0.81 | 0.86 | 1.07 | 0.88 | 1.00 | 1.63 | 23.99 | 23.20 | 23.54 | 25.91 | 23.42 | 24.24 | 556,877 | 578,372 |
| 1995 | 1.49 | 1.61 | 1.54 | 1.48 | 1.63 | 2.12 | 26.46 | 26.40 | 28.07 | 27.63 | 27.17 | 28.21 | 556,877 | 578,372 |
| 1995 | 0.90 | 1.60 | 0.88 | 1.04 | 1.18 | 1.68 | 23.98 | 23.06 | 29.63 | 26.76 | 24.74 | 24.82 | 556,877 | 578,372 |
| 1995 | 1.01 | 1.13 | 1.00 | 1.18 | 1.47 | 1.99 | | | | | | | 556,877 | 578,372 |
| 1995 | 0.98 | 1.01 | 1.22 | 1.03 | 1.24 | 1.77 | 24.97 | 24.13 | 23.66 | 26.86 | 22.66 | 23.34 | 556,877 | 578,372 |
| 1995 | 1.04 | 1.03 | 0.92 | 1.04 | 1.24 | 1.82 | 24.46 | 23.70 | 28.58 | 26.51 | 26.08 | 27.40 | 556,877 | 578,372 |
| 1995 | 0.97 | 1.15 | 1.10 | 1.44 | 1.59 | 2.19 | 25.77 | 25.65 | 27.11 | 25.44 | 24.86 | 26.30 | 556,877 | 578,372 |
| 1995 | 1.10 | 1.19 | 1.13 | 1.25 | 1.38 | 2.14 | 22.81 | 22.73 | 26.99 | 25.59 | 24.10 | 24.79 | 556,877 | 578,372 |
| 1995 | 1.43 | 1.60 | 1.53 | 1.66 | 1.74 | 2.30 | 26.26 | 25.72 | 26.28 | 27.66 | 26.16 | 26.50 | 556,877 | 578,372 |
| 1995 | 1.39 | 1.45 | 1.43 | 1.51 | 1.59 | 1.99 | 24.70 | 24.48 | 30.52 | 24.54 | 24.19 | 25.64 | 556,877 | 578,372 |
| 1995 | 1.53 | 1.63 | 1.84 | 1.62 | 1.79 | 2.05 | 25.04 | 23.92 | 25.88 | 25.17 | 23.80 | 25.22 | 556,877 | 578,372 |
| 1995 | 1.48 | 1.46 | 1.62 | 1.50 | 1.66 | 2.13 | 23.45 | 23.41 | 29.00 | 25.10 | 24.77 | 25.52 | 556,877 | 578,372 |
| 1995 | 1.31 | 1.43 | 1.44 | 1.66 | 1.74 | 2.11 | | | | | | | 556,877 | 578,372 |
| 1995 | 1.68 | 1.82 | 1.98 | 2.32 | 2.38 | 2.53 | 25.53 | 25.10 | 30.30 | 26.35 | 24.28 | 25.46 | 556,877 | 578,372 |
| 1995 | 1.48 | 1.60 | 1.64 | 1.75 | 1.94 | 2.58 | 22.94 | 22.82 | 31.87 | 24.86 | 23.98 | 24.56 | 556,877 | 578,372 |

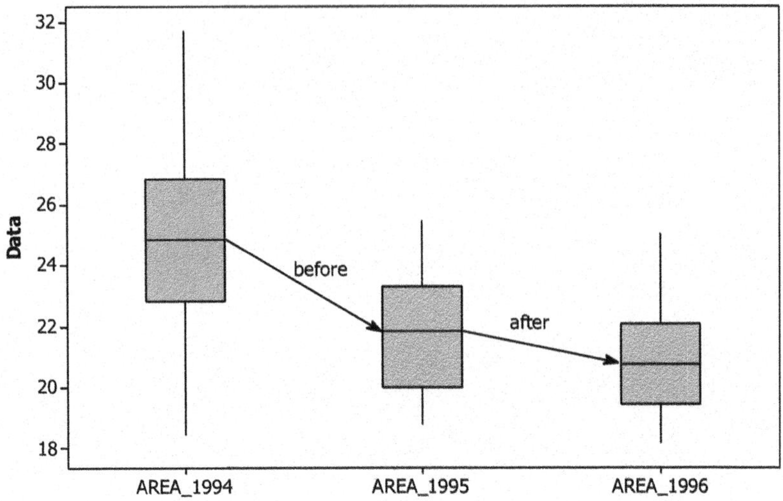


**Fig. 4.9** Box plot of Area deflection basin parameter (pavement strength), before and after

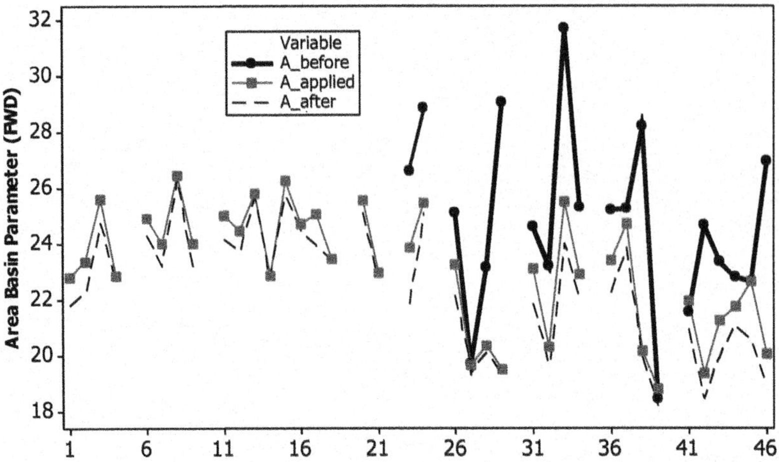


**Fig. 4.10** Profile of structure condition (Area deflection basin) before and after a surface treatment

A similar line trend for group 2 confirmed that the rate of deterioration on pavement structure slowed down for the majority of segments (Fig. 4.10). It can also be observed how closely the year-applied and year-after lines are in contrast with the significant drop from the year-before line, for both groups, demonstrating that the application of a surface treatment seems to have an immediate effect of slowing down pavement structural degradation.

### 4.3.4 Capturing Treatment Effectiveness

A Markov chain was used to capture before and after treatment effectiveness for the dataset. The main idea was to observe the change of IRI for a number of segments and to express such a change on a probability matrix, in which the distance from the main diagonal represents the extension of service life. Therefore, synchronizing the values on the matrix with annual deterioration rates such that the distance between two cells corresponded to a fixed number of years (preferably one or two).

According to Opus (2006), roughness progression of AC arterial roads in New Brunswick follows an exponential relationship (Eq. 4.3). Therefore, an apparent age related to the condition was obtained from Eq. 4.3 starting at IRI = 1 m/km (for age = zero) and progressing to age 20. Eleven clusters ranging from 1 m/km to 2.5 m/km were defined using previously obtained ages and corresponding IRI as threshold values. A count of segments moving across clusters was used to capture treatment effectiveness in a transition probability matrix.

$$y = e^{0.0424x} \tag{4.3}$$

Each cell movement corresponded to 2 years of apparent age. Three segments were eliminated from the original database as they showed decay. The main diagonal corresponded to the likelihood of a segment receiving treatment and gaining less than 2 years in lifespan extension. As observed, pavements in good (less than 1.4 m/km IRI) condition did not benefit from surface treatments. Segments with IRI values above 1.66 and receiving a surface treatment seem to gain between 6 and 10 years of additional life. Pavement with IRI values between 1.4 and 1.53 seems to only gain 2 years of lifespan extension (Table 4.11). IRI reset value after receiving a surface treatment lies between 1.18 and 1.29 m/km.

Figure 4.11 illustrates effectiveness for those segments with $1.81 < \text{IRI} < 1.97$ receiving a surface treatment; as seen in Table 4.11, 67% of such segments will gain a lifespan extension of 6 years while 33% will gain up to 8 years.

Deterioration curves for Area deflection basin parameters were used to synchronize pavement structure capacity with apparent age. Traffic intensity did not seem to affect the rate of deterioration of pavement structure. A best-fit curve based on the apparent age approach was used to estimate the rate of deterioration as shown in Eq. 4.4.

$$y = -0.0141x^3 + 0.2767x^2 - 2.7998x + 28.477 \tag{4.4}$$

A similar transition probability matrix was developed for the before (1994–1995) and after trend (1995–1996); the idea was to measure differences in deterioration rate in terms of age (years). Tables 4.12 and 4.13 show such TPM. The decay of pavement structural capacity before the application of surface treatment is much faster than after having received a surface treatment. The before trends indicate a majority of segments decaying between 2 and 4 apparent ages as measured by the deterioration model (Table 4.12). Decay rates slowed to about 1 year for all of the segments (Table 4.13).

**Table 4.11** Transition probability matrix of surface treatment effectiveness

| IRI range | Age | 1.00–1.09 | 1.09–1.18 | 1.18–1.29 | 1.29–1.40 | 1.40–1.53 | 1.53–1.66 | 1.66–1.81 | 1.81–1.97 | 1.97–2.15 | 2.15–2.33 | 2.33–2.54 |
|---|---|---|---|---|---|---|---|---|---|---|---|---|
| 1.00–1.09 | 0 | 0 | – | – | – | – | – | – | – | – | – | – |
| 1.09–1.18 | 2 | 0 | 0 | – | – | – | – | – | – | – | – | – |
| 1.18–1.29 | 4 | 0 | 0 | 0 | – | – | – | – | – | – | – | – |
| 1.29–1.4 | 6 | 0 | 0 | 0 | 100% | – | – | – | – | – | – | – |
| 1.4–1.53 | 8 | 0 | 0 | 0 | 100% | 0 | – | – | – | – | – | – |
| 1.53–1.66 | 10 | 0 | 0 | 17% | 17% | 33% | 33% | – | – | – | – | – |
| 1.66–1.81 | 12 | 0 | 0 | 100% | 0 | 0 | 0 | 0 | – | – | – | – |
| 1.81–1.97 | 14 | 0 | 0 | 0 | 33% | 67% | 0 | 0 | 0 | – | – | – |
| 1.97–2.15 | 16 | 0 | 0 | 25% | 0 | 50% | 25% | 0 | 0 | 0 | – | – |
| 2.15–2.33 | 18 | 0 | 0 | 0 | 50% | 0 | 0 | 50% | 0 | 0 | 0 | – |
| 2.33–2.54 | 20 | 0 | 0 | 0 | 0 | 100% | 0 | 0 | 0 | 0 | 0 | 0 |

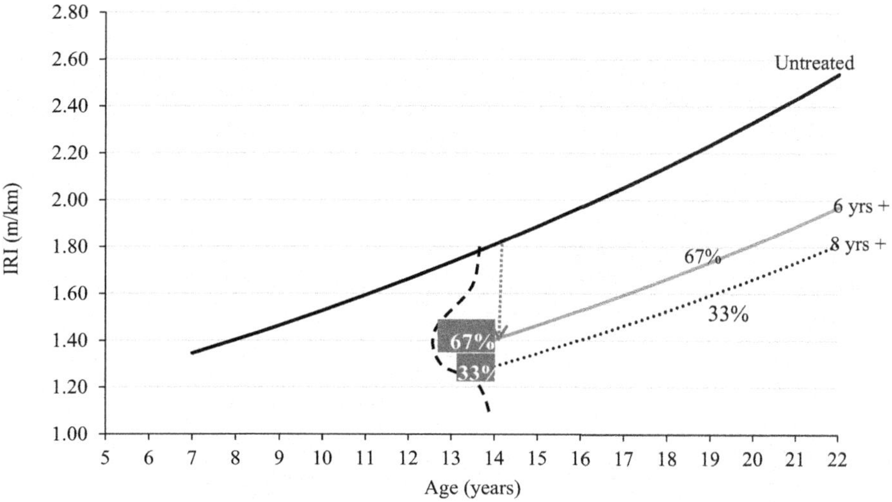


**Fig. 4.11** Sample lifespan extension for treated segments at 1.81 < IRI < 1.97

**Table 4.12** Area deflection basin—Pavement structure deterioration: before surface treatment

| Age | Area | 28.48–32.6 | 25.94–28.47 | 23.87–25.93 | 22.19–23.86 | 20.80–22.18 | 19.64–20.79 | 18.60–19.63 | 17.61–18.59 |
|---|---|---|---|---|---|---|---|---|---|
| 0 | 28.48–32.6 | | 0 | 67% | 0 | 0 | 0 | 33% | 0 |
| 1 | 25.94–28.47 | - | | 0 | 33% | 0 | 67% | 0 | 0 |
| 2 | 23.87–25.93 | - | - | | 83% | 0 | 0 | 17% | 0 |
| 3 | 22.19–23.86 | - | - | - | | 40% | 40% | 0 | 0 |
| 4 | 20.80–22.18 | - | - | - | - | | 0 | 0 | 0 |
| 5 | 19.64–20.79 | - | - | - | - | - | | 100% | 0 |
| 6 | 18.60–19.63 | - | - | - | - | - | - | | 0 |
| 7 | 17.61–18.59 | - | - | - | - | - | - | - | |

**Table 4.13** Area deflection basin – Pavement structure deterioration: after surface treatment

| Age | Area | 28.48–32.6 | 25.94–28.47 | 23.87–25.93 | 22.19–23.86 | 20.80–22.18 | 19.64–20.79 | 18.60–19.63 | 17.61–18.59 |
|---|---|---|---|---|---|---|---|---|---|
| 0 | 28.48–32.6 | | 0 | 0 | 0 | 0 | 0 | 0 | 0 |
| 1 | 25.94–28.47 | - | | 0 | 0 | 0 | 0 | 0 | 0 |
| 2 | 23.87–25.93 | - | - | | 33% | 0 | 0 | 0 | 0 |
| 3 | 22.19–23.86 | - | - | - | | 67% | 17% | 0 | 0 |
| 4 | 20.80–22.18 | - | - | - | - | | 33% | 0 | 0 |
| 5 | 19.64–20.79 | - | - | - | - | - | | 25% | 0 |
| 6 | 18.60–19.63 | - | - | - | - | - | - | | 50% |
| 7 | 17.61–18.59 | - | - | - | - | - | - | - | |

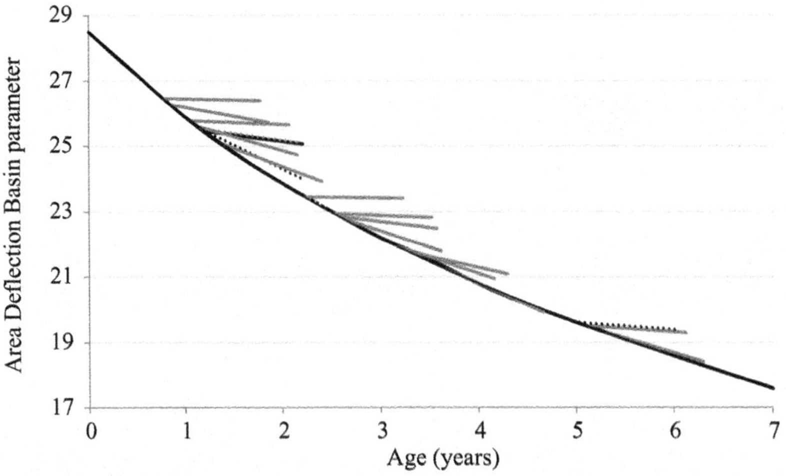


**Fig. 4.12** Observed deterioration trends for structural capacity after surface treatment

Figure 4.12 shows the original deterioration curve and detailed trends for selected segments. A reduction in the rate of decay of structural capacity can be observed. Unfortunately, all available segments were treated after 2 years of receiving the surface treatment, impeding to observe longer after-treatment trends as those reported by other researchers.

## 4.4 Exercises

**Exercise 4.1**
A company observed before and after treatment for reconstruction in different years and results are presented in Table 4.14.

A. What is the operational window and effectiveness for reconstruction?

| Intervention | From (IRI) | To (IRI) | Gain (IRI) |
|---|---|---|---|
| Reconstruction | | | |

**Exercise 4.2**
A city did a condition assessment for the wastewater pipeline network based on a 1–10 performance indicator. The assessment was done in 2019 and 2020 for 40 pipes as below. Between these two assessments, pipes partially received a preventive (for pipe in condition range 6–8) or a corrective (for pipes in condition range of 4–5) maintenance. What is the effectiveness of these two interventions in both condition grade and age?

Hint: You first need to develop a deterioration model using data.

**Exercise 4.3**
The Karan Company recently has built a 100 km highway for $220,000 per kilometer. Karan also is responsible to maintain it for 20 years. They assessed pavement performance (IRI) in January 2017 and 2019. Figure 4.13 shows the deterioration model (i.e., curve) for this highway. Immediately after the first observation (the year 2017), they applied different maintenance actions including two new technologies to overlay the highway. The recommended trigger level for overlay type 1 was $1.5 < \text{IRI} < 2.0$ (m/km) and type 2 was $2 < \text{IRI} < 2.5$ (m/km). The cost of overlay type 1 per kilometer is $85,000 and type 2 is $100,000. To evaluate the effectiveness (age) of these new technologies, the company used the below developed Markov probability matrix (Table 4.15). Typically, they do reconstruction when segment IRI (m/km) passes 3.

**Table 4.14** Exercise 4.1 dataset

| Intervention | IRI before treatment | Year | IRI after treatment |
|---|---|---|---|
| Reconstruction | 3 | 2000 | 0.5 |
| Reconstruction | 2.9 | 2005 | 0.7 |
| Reconstruction | 3 | 2008 | 0.55 |
| Reconstruction | 2.9 | 2010 | 0.65 |
| Reconstruction | 3.1 | 2011 | 0.63 |
| Reconstruction | 2.95 | 2013 | 0.75 |
| Reconstruction | 3.2 | 2015 | 0.54 |
| Reconstruction | 2.9 | 2016 | 0.58 |

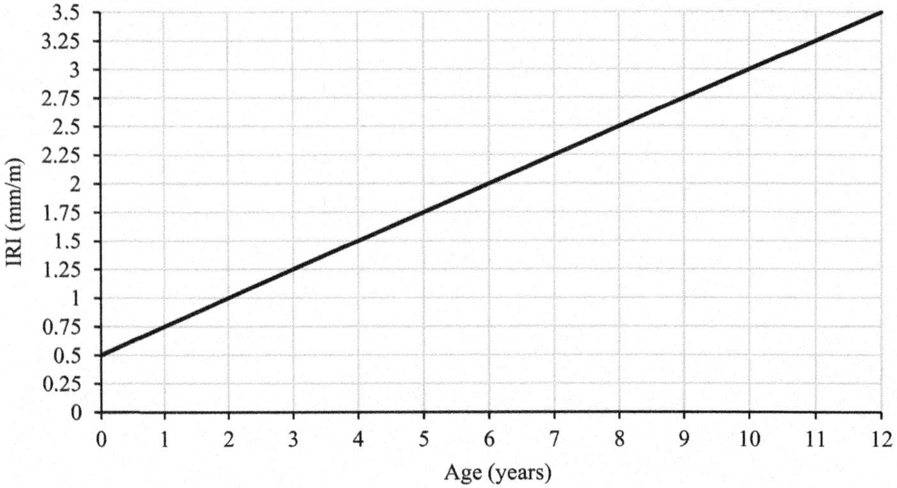


**Fig. 4.13** Deterioration model for Exercise 4.3

**Table 4.15** Markov matrix for treatments effectiveness

| Age | | | | | | | | | | | |
|---|---|---|---|---|---|---|---|---|---|---|---|
| 0 | 0 | – | – | – | – | – | – | – | – | – | – |
| 1 | 0 | 0 | – | – | – | – | – | – | – | – | – |
| 2 | 0 | 80% | 20% | – | – | – | – | – | – | – | – |
| 3 | 0 | 0 | 0 | 0 | – | – | – | – | – | – | – |
| 4 | 5% | 85% | 10% | 0 | 0 | – | – | – | – | – | – |
| 5 | 0 | 5% | 85% | 10% | 0 | 0 | – | – | – | – | – |
| 6 | 0 | 0 | 0 | 0 | 0 | 0 | 0 | – | – | – | – |
| 7 | 0 | 0 | 5% | 40% | 35% | 15% | 5% | 0 | – | – | – |
| 8 | 0 | 0 | 0 | 0 | 0 | 0 | 0 | 0 | 0 | – | – |
| 9 | 0 | 0 | 0 | 0 | 0 | 0 | 0 | 0 | 0 | 0 | – |
| 10 | 100% | 0 | 0 | 0 | 0 | 0 | 0 | 0 | 0 | 0 | 0 |

A. What would be the effectiveness (in the number of years of lifespan extension) for these two methods?
B. Which mill and overlay method (type 1 or 2) do you recommend to this company? Why?

# References

Amador Jimenez, L., & Mrawira, D. (2011). Capturing variability in pavement performance models from sufficient time-series predictors: A case study of the New Brunswick road network. *Canadian Journal of Civil Engineering, 38*(2), 210–220.

Opus. New Brunswick Department of transportation. (2006). *Asset management business framework project – Pavement deterioration curves and performance measurement. Report FR-TAM-06-01*. Opus International Consultants.

Xu, B., Ranjithan, S. R., & Kim, Y. R. (2003). New relationships between falling weight deflectometer and asphalt pavement layer condition indicators. Transportation Research Record: Journal of the Transportation Research Board, No. 1806, Transportation Research Board of the National Academies, Washington, D.C., pp. 48–56.

# Chapter 5
# Decision Support Systems for IAM

## 5.1 Decision Levels: Strategic, Tactical, and Operational

Federal, state, provincial, territorial, and municipal governments and all other asset owners around the world must make decisions to invest public funds to maintain, rehabilitate, upgrade, and expand public infrastructure while budget limitation brings a need for seeking the most cost-effective and optimum solutions. Thus, decision-making is an essential step of this journey for most asset owners, and mature organizations in AM always are seeking optimum solutions through advanced tools and techniques.

Decisions trigger the need for plans where alternatives are compared and shortlisted. Selected alternatives become the basis for the budget, which is most commonly done on an annual basis. This level of planning and decisions is called operational, and it accompanies annual planning, but it is unable to consider the impact of decisions in the long term (also known as the strategic level).

Mature organizations in AM need to be proactive, and having long-term plans is required to avoid surprise and plan ahead of failure. As was discussed earlier, lack of data and uncertainty are always challenging in this kind of planning; however, it should not stop developing such strategic renewal planning as that is the nature of projecting the future.

The asset management team must put enough effort and use the best solutions in the literature to build the most accurate forecasting models; however, a tactical (medium-term) plan always gives a chance to adjust the strategic plan where is required and recover uncertainty. For example, if the model says a specific asset should be replaced in the next 3 years while recent observation indicates an emergency replacement is required or vice versa. The operation team should use a 3–5 year window in a tactical plan to address these sorts of changes.

Alternatives are often compared in terms of the return on investment, and if more mature, in terms of future maintenance needs during their lifecycles, all the way to the point on time when replacement is needed. This is triggered by valuation



A. Mohammadi, L. Amador Jimenez, *Asset Management Decision-Making For Infrastructure Systems*, https://doi.org/10.1007/978-3-030-97614-9_5

methods enforced by accounting laws at public organizations, which require financial analysis of any investment, accompanied by depreciation methods and estimation of salvage values. This financial provision needs to be accompanied by an engineering analysis that enables tracking KPIs interrelated to the selected alternative that will change over time as a consequence of the new infrastructure or facility being funded (through maintenance, rehabilitation, upgrade, or expansion). Such long term analysis is called strategic; it allows to have a global view of the infrastructure systems and their interactions; and it enables the forecasting of future levels of performance indicators, some financial, but most relevant to the ability to service the public, support economic growth, provide social well-being, and protect the environment.

However, decisions on new infrastructure had also repercussions on the demand and often affect other systems, which is most likely neglected in current classical AM practice and should be captured in planning for key infrastructure such as urban transit systems. For example, consider the construction of a new railway transit line that will serve several existing towns and villages connecting them to the rest of the rail network and the city. Not only does the new railway require maintenance after being opened but also it will affect the modal distribution of other existing transportation options such as cars and buses using roads; therefore, the level of congestion on the road links used by those villages and towns to travel to the city may experience a reduction on the number of vehicles per hour during the peak hour, resulting in an improvement on the level of congestion and shortened travel time. The reduced number of vehicles could result also in a reduction of road collisions at the intersections used before by work commuters. If the new railway line is used at night time to move cargo (goods and merchandises) between the city and the so mentioned towns and villages, then the main roads currently used by trucks to move the same will experience less traffic loading and their rate of deterioration will be reduced. Let us think the new railway comes accompanied by park-and-ride, drop-and-kiss facilities at all new stations: all intersections leading to the stations might likely observe an increase in the number of vehicles during the peak time (and likely, but not as severe, at other peaks but because of the stratified work-exit-time), associated to an increase in the level of congestion and a drop in the level of safety of road users. In a nutshell, any new piece of infrastructure must be assessed in a multi-systems context to attempt to clearly capture its benefits and drawbacks in the short, medium, and long terms.

The need for a 3–5 years tactical planning comes from the fact that long-term plans are always followed with a degree of uncertainty in deterioration, intervention effectiveness, as well as constraints such as available budget. This brings a need to review coming years incorporating the most recent information, which may reprioritize actions.

Furthermore, most decision support systems provide a prioritized list of projects, which often disregards any measures of spatial and temporal proximity between the selected actions to be implemented. For example, think of the maintenance of a pavement, which requires mill and overlay. Now imagine there are multiple sections of the road schedule next year and the year after and they are one next to the other. In such a case, keeping the list of maintenance projects as is will lead to traffic disruptions next year and the year after. Perhaps merging both either this year or next year is better as traffic will be disrupted only once. The disruption of traffic has impacts on the safety of motorists and other road users (pedestrians, cyclists); every time that maintenance repairs occur, a temporary traffic management plan needs to be set in place to prevent road accidents and manage the detours of traffic. The disruption of traffic produces longer travel times and affects the productivity of labor in the region (people idling in traffic) and increases the cost structure of goods being moved (freight traveling). In addition, longer travel time implies more fuel and more gas emissions, having a detriment to the environment. Therefore, merging various segments where maintenance was originally scheduled for next year and the following year resulted in advantages in terms of safety and disruptions to the users (Mohammadi et al., 2020).

Besides, bundling these projects reduce construction cost and drop wastes such as multiple mobilizations and demobilizations. Bundling the maintenance of various segments together also aid in securing contractors from other nearby locations because the magnitude of works increases and makes profitable the participation of contractors located farther away. Having larger packages of works could reduce the cost quoted and bring maintenance services otherwise not available (hot in-place recycle per instance) when the package of works increases (Mohammadi et al., 2020).

From a political standpoint, most governments step into power for a given period of time, commonly 4–5 years. Therefore, having a tactical plan helps to have a concrete target for what is accomplishable within such a period. However, it is discouraged for governments to attempt to change tactical plans, as those plans come from long-term planning as explained before.

From an IAM perspective, a mature platform first plans for a long-term (i.e., strategic), which can be 10, 20, or even longer years depending on the type of infrastructure to ensure the long impacts of current decisions are captured and goals will be achieved. Then, the next 3–5 years will be focused to extract tactical plans, based on which actions are later prioritized annually (i.e., operational).

## 5.2 Decision-Making Approaches: Ranking, Prioritization, Optimization

Every agency prepares a budget to preserve its assets, and always some sort of planning precedes budgeting. The quality of the planning and the budgeting process have a major impact on the condition of the asset network and on the lifecycle cost of maintaining the assets. This section explains the several approaches used to select investments to preserve infrastructure assets. We kick off with the simplest one (ranking) and increase the level of sophistication until we arrive at optimization.

## 5.2.1 Ranking

The ranking (also known as worst-first) is the simplest approach; it is usually performed in accordance with agency guidelines where treatments are selected to match current condition (a weighting factor may be used to create a condition index) and alternatives for a given year are sorted (ranked) and selected until all the available budget is spent.

A typical ranking will obey the following steps:

- Step 1. The agency assesses the level of condition for all infrastructure assets in its jurisdiction.
- Step 2. Agency rules are used to assign a matching treatment according to the results of Step 1.
- Step 3. A cost estimate is prepared for each asset.
- Step 4. Choices are sorted using given criteria: this can go as sophisticated as cost-benefit ratios to as simple as the worst-first approach (worst condition first, best condition last).
- Step 5. Choices are selected until the available budget for the given year is spent.

**Example 5.1**

Let's start with a simple case where some storm pipes are assessed and Table 5.1 provides the condition level, assigned intervention based on their current condition, and the corresponding cost.

Assuming an available budget of $10 million for the coming year, in a basic ranking approach, pipes are ranked based on their condition levels as Table 5.2. Then, respecting the available budget, the first three pipes can be maintained in the first year.

This approach can be improved by assigning a weighting factor to each alternative. For example, the criticality of each pipe (e.g., depending on where it is located in the network) can be assigned a 0–1 weight (less means more critical) as described in Table 5.3.

**Table 5.1** Example 5.1 dataset

| Pipe ID | Condition level (%) | Treatment | Cost ($M) |
|---|---|---|---|
| 16A | 38 | Major maintenance | 3.5 |
| 18C | 45 | Major maintenance | 2.5 |
| 3F | 49 | Minor maintenance | 1.25 |
| 15A | 55 | Minor maintenance | 2 |
| 4B | 75 | Preventive maintenance | 0.6 |
| 10C | 80 | Preventive maintenance | 0.5 |
| 8B | 83 | Preventive maintenance | 0.55 |
| 5A | 40 | Major maintenance | 4 |
| 6D | 60 | Minor maintenance | 1.5 |
| 7D | 66 | Minor maintenance | 1.75 |

**Table 5.2** Example 5.1 solution

| Pipe ID | Ranking | Condition level (%) | Treatment | Cost ($M) |
|---|---|---|---|---|
| 16A | 1 | 38 | Major maintenance | 3.5 |
| 5A | 2 | 40 | Major maintenance | 4 |
| 18C | 3 | 45 | Major maintenance | 2.5 |
| 3F | 4 | 49 | Minor maintenance | 1.25 |
| 15A | 5 | 55 | Minor maintenance | 2 |
| 6D | 6 | 60 | Minor maintenance | 1.5 |
| 7D | 7 | 66 | Minor maintenance | 1.75 |
| 4B | 8 | 75 | Preventive maintenance | 0.6 |
| 10C | 9 | 80 | Preventive maintenance | 0.5 |
| 8B | 10 | 83 | Preventive maintenance | 0.55 |

**Table 5.3** Example 5.1 solution (2)

| Pipe ID | Condition level (%) | Weight (Criticality) | Treatment | Cost ($M) |
|---|---|---|---|---|
| 16A | 38 | 0.8 | Major maintenance | 3.5 |
| 5A | 40 | 0.35 | Major maintenance | 4 |
| 18C | 45 | 0.55 | Major maintenance | 2.5 |
| 3F | 49 | 0.95 | Minor maintenance | 1.25 |
| 15A | 55 | 1 | Minor maintenance | 2 |
| 6D | 60 | 0.65 | Minor maintenance | 1.5 |
| 7D | 66 | 0.66 | Minor maintenance | 1.75 |
| 4B | 75 | 0.4 | Preventive maintenance | 0.6 |
| 10C | 80 | 0.35 | Preventive maintenance | 0.5 |
| 8B | 83 | 0.76 | Preventive maintenance | 0.55 |

For the same example.

In this case, instead of ranking by only condition level, the ranking factor would be a combination (i.e., multiplying) of condition level and weight (Table 5.4).

Then by ranking pipes based on (condition × criticality), a new list of prioritized pipes will be identified, which is different than Table 5.5.

The ranking approach still can be enhanced by considering multiple condition indicators, which is explained in this coming example.

**Example 5.2**

Consider the maintenance for pavements of six road sections on a small municipality with a budget of USD$ 5,000,000 per year. The six segments and their current level of damage and structural capacity are presented in Table 5.6.

For this example, a set of rules to select the most appropriate treatment are based on detailed damage indicators of average rutting (mm), average cracked area (%), and deflection basin area (DBA) scaled from 11 (weak) to 36 (strong). The agency has the following rules to judge on the applicable treatments:

**Table 5.4** Example 5.1 solution (3)

| Pipe ID | Condition level (%) | Weight (Criticality) | Condition*criticality | Treatment | Cost ($M) |
|---|---|---|---|---|---|
| 16A | 38 | 0.8 | 30.4 | Major maintenance | 3.5 |
| 5A | 40 | 0.35 | 14 | Major maintenance | 4 |
| 18C | 45 | 0.55 | 24.75 | Major maintenance | 2.5 |
| 3F | 49 | 0.95 | 46.55 | Minor maintenance | 1.25 |
| 15A | 55 | 1 | 55 | Minor maintenance | 2 |
| 6D | 60 | 0.65 | 39 | Minor maintenance | 1.5 |
| 7D | 66 | 0.66 | 43.56 | Minor maintenance | 1.75 |
| 4B | 75 | 0.4 | 30 | Preventive maintenance | 0.6 |
| 10C | 80 | 0.35 | 28 | Preventive maintenance | 0.5 |
| 8B | 83 | 0.76 | 63.08 | Preventive maintenance | 0.55 |

**Table 5.5** Example 5.1 solution (4)

| Pipe ID | Ranking | Condition level (%) | Weight (Criticality) | Condition*criticality | Treatment | Cost ($M) |
|---|---|---|---|---|---|---|
| 5A | 1 | 40 | 0.35 | 14 | Major maintenance | 4 |
| 18C | 2 | 45 | 0.55 | 24.75 | Major maintenance | 2.5 |
| 10C | 3 | 80 | 0.35 | 28 | Preventive maintenance | 0.5 |
| 4B | 4 | 75 | 0.4 | 30 | Preventive maintenance | 0.6 |
| 16A | 5 | 38 | 0.8 | 30.4 | Major maintenance | 3.5 |
| 6D | 6 | 60 | 0.65 | 39 | Minor maintenance | 1.5 |
| 7D | 7 | 66 | 0.66 | 43.56 | Minor maintenance | 1.75 |
| 3F | 8 | 49 | 0.95 | 46.55 | Minor maintenance | 1.25 |
| 15A | 9 | 55 | 1 | 55 | Minor maintenance | 2 |
| 8B | 10 | 83 | 0.76 | 63.08 | Preventive maintenance | 0.55 |

**Table 5.6** Full network of pavement sections

| Section | Average rutting (mm) | Average cracked area (%) | Deflection basin area (11 min, 36 max) | PCI | Best treatment | Total cost (USD$ million) |
|---|---|---|---|---|---|---|
| Main St. Sect.1 | 11.5 | 4 | 30 | 86 | Microsurface | 0.5 |
| McGill St. Sect.4 | 20 | 15 | 34 | 84 | Mill and overlay | 1 |
| Spence Av. Sect.2 | 25 | 30 | 14 | 27 | Full reconstruction | 3 |
| Edmont St. Sect.5 | 20 | 30 | 21 | 47 | Partial reconstruction | 2 |
| Edmon St. Sect.8 | 2 | 8 | 29 | 85 | Crack sealing | 0.25 |
| Stevens St. Sect. 7 | 15 | 18 | 24 | 66 | Mill and overlay | 1 |

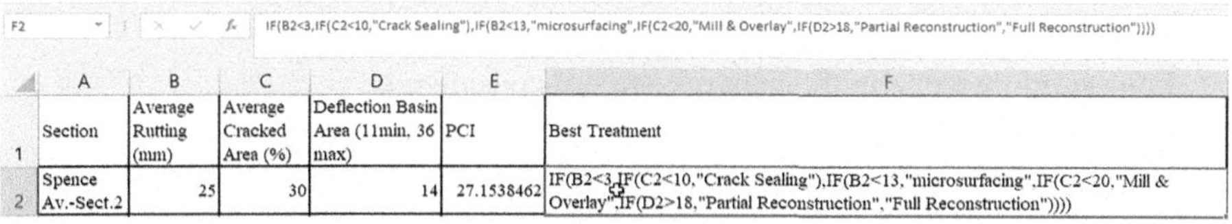
F2 IF(B2<3,IF(C2<10,"Crack Sealing"),IF(B2<13,"microsurfacing",IF(C2<20,"Mill & Overlay",IF(D2>18,"Partial Reconstruction","Full Reconstruction"))))

| | A | B | C | D | E | F |
|---|---|---|---|---|---|---|
| 1 | Section | Average Rutting (mm) | Average Cracked Area (%) | Deflection Basin Area (11min. 36 max) | PCI | Best Treatment |
| 2 | Spence Av.-Sect.2 | 25 | 30 | 14 | 27.1538462 | IF(B2<3,IF(C2<10,"Crack Sealing"),IF(B2<13,"microsurfacing",IF(C2<20,"Mill & Overlay",IF(D2>18,"Partial Reconstruction","Full Reconstruction")))) |

**Fig. 5.1** Example of decision rule implemented on MS-excel

- Rutting <3 mm and cracked area <10% do crack sealing
- 3 mm < Rutting <13 mm can receive micro surfacing
- Rutting >13 mm check cracked area, if less than 20% segment can receive mill and overlay of HMA layer
- Rutting >13, cracked area > 20%, DBA > 18, partial reconstruction (new HMA + base re-compaction and sealing)
- Rutting >13 and cracked area >20%, if DBA <18 do full reconstruction

These rules can be built into an MS-Excel (or any other platform) to automate the decision making as illustrated in Fig. 5.1.

PCI is used to guide ranking as criteria and is formed from the damage indicators using an equal weight of surface defects and structural capacity with the following equation:

$$\text{PCI} = \{0.5^* [100 - (\text{Rutting}/13)^* \text{ Cracked area}\%]\} + 0.5^* [100^* (\text{DBA} - 11)/25] \quad (5.1)$$

Ranked options are presented in Table 5.7 and the first two segments are selected for maintenance this year, leaving all other segments for future periods.

The problem with ranking is that it neglects several aspects important to the management and scheduling of investments:

**Table 5.7** Example of ranking for pavement sections—Worst first ranked options

| Section | Average rutting (mm) | Average cracked area (%) | Deflection basin area (11 min, 36 max) | PCI | Best treatment | Total cost (USD$ millions) |
|---|---|---|---|---|---|---|
| Spence Av. Sect.2 | 25 | 30 | 14 | 27 | Full reconstruction | 3 |
| Edmont St. Sect.5 | 20 | 30 | 21 | 47 | Partial reconstruction | 2 |
| Stevens St. Sect. 7 | 15 | 18 | 24 | 66 | Mill and overlay | 1 |
| McGill St. Sect.4 | 20 | 15 | 34 | 84 | Mill and overlay | 1 |
| Edmon St. Sect.8 | 2 | 8 | 29 | 85 | Crack sealing | 0.25 |
| Main St. Sect.1 | 11.5 | 4 | 30 | 86 | Microsurface | 0.5 |

1. The rate of deterioration is not considered.
2. Long-term impacts on the network are not considered.
3. Economic analysis for alternative strategies is not considered.
4. Gained efficiencies by full optimization of resources are not considered.

**Example 5.3**
Let us look at how considering the rate of deterioration changes in investment planning. Let us continue with the same example; imagine that the rate of deterioration of the sections is known. This allows us to forecast the condition next year, the required treatment (next year), and its associated cost as shown in Table 5.8. As seen on the table, the first two road segments originally scheduled to be reconstructed (Spence Av. Sect.2 and Edmont St. Sect.5) are still candidates for the same investments.

However, all other sections have further deteriorated and are now candidates for more expensive investments, for instance, Main St. moved from micro-surfacing (at 0.5 USD million) to mill and overlay at 1 USD million.

From an economic analysis perspective, the worst first strategy requires 5 million USD in 2021 and 5.5 million in the year 2022.

Now we ask: What would have been the case if we start by preserving the roads in better condition before fixing the ones in poor condition (called good first approach)?

Table 5.9 shows the case for the assumed 5 million USD budget. As seen, all segments except Spence Av. Sect.2 could have received a treatment that would have rejuvenated them leaving us with a need for 3 million USD for the year 2021, instead of the 5.5 million USD of the worst first approach. A net saving of 2.5 million dollars or 50% of the annual budget.

This example clearly shows the limitation of ignoring deterioration. The ranking is a limited approach that prioritizes investments for 1 year, and hence is the equivalent of a one-year prioritization.

**Table 5.8** Lack of consideration of deterioration rate on ranking

| Section | 2021 average rutting (mm) | 2021 average cracked area (%) | 2021 deflection basin area (11 min, 36 max) | 2021 PCI | 2021 best treatment | 2022 total cost (USD$ millions) | 2022 average rutting (mm) | 2022 average cracked area (%) | 2022 deflection basin area (11 min, 36 max) | 2022 PCI | 2022 best treatment | 2022 total cost (USD$ millions) |
|---|---|---|---|---|---|---|---|---|---|---|---|---|
| Spence Av. Sect.2 | 25 | 30 | 14 | 27 | Full reconstruction | 3 | | | | | | |
| Edmont St. Sect.5 | 20 | 30 | 21 | 47 | Partial reconstruction | 2 | | | | | | |
| Stevens St. Sect. 7 | 15 | 18 | 24 | 66 | Mill and overlay | 1 | 18 | 20 | 21 | 56.15 | Partial reconstruction | 2 |
| McGill St Sect.4 | 20 | 15 | 34 | 84 | Mill and overlay | 1 | 23 | 20 | 33 | 76.31 | Partial reconstruction | 2 |
| Edmon St. Sect.8 | 2 | 8 | 29 | 85 | Crack sealing | 0.25 | 6 | 12 | 27.5 | 80.23 | Microsurfacing | 0.5 |
| Main St. Sect.1 | 11.5 | 4 | 30 | 86 | Microsurfacing | 0.5 | 15 | 7 | 28.5 | 80.96 | Mill and overlay | 1 |

Estimated values for 2022 are based on Engineer's criteria

**Table 5.9** Good first approach

| Section | Average rutting (mm) : Before / After | Average cracked area (%): Before / After | Deflection basin area (11min, 36 max): Before / After | PCI: Before / After | Best treatment | Total Cost (USD$ millions) |
|---|---|---|---|---|---|---|
| Main St.-Sect.1 | 8 / 0 | 5 / 0 | 30 / 30 | 86 | Microsurfacing | 0.5 |
| Edmon St. Sect.8 | 2 / 2 | 8 / 0 | 29 / 29 | 85 | Crack sealing | 0.25 |
| McGill St.-Sect.4 | 20 / 0 | 15 / 0 | 34 / 34 | 84 | Mill and overlay | 1 |
| Stevens St. Sect. 7 | 15 / 0 | 18 / 0 | 24 / 24 | 66 | Mill and overlay | 1 |
| Edmont St. Sect.5 | 20 / 0 | 30 / 0 | 21 / 34 | 47 | Partial reconstruction | 2 |
| Spence Av.-Sect.2 | 25 | 30 | 14 | 27 | Full reconstruction | 3 |

Note: Shaded cells represent road segments that will receive treatment

#### 5.2.1.1 Risk-Based Ranking

Applying a risk assessment to prioritize assets is also a common method to improve ranking where the highest rank is given to an SGR project with the highest measured risk score. Risk management is an extensive topic, but for this purpose, the decision-maker needs to evaluate the risk score (RS) for each asset/system and its required SGR project by estimating two main components of the probability of failure (P) and consequence of failure (C) (Eq. 5.2). By including the consequences (e.g., safety, security, reliability, and availability) of any interruption in the service, this method is seeking a performance-based approach enhancing ranking.

$$\mathrm{RS} = P \times C \tag{5.2}$$

Probability of Failure

The probability of failure for an asset or a system can be estimated using Eq. 5.3 linking failure ($F$) to asset performance (i.e., reliability [$R$]) as was described in Sect. 2.3.6.1. P can be defined in a 0–100 or 0–1 range.

$$P = F = 1 - R \tag{5.3}$$

Condition assessment (e.g. 1–5 scale) can be scaled to apply here and measure $P$. For instance, a grade 1 can be interoperated to 20% reliability and therefore 80% chance of failure.

Consequence of Failure

It would be more challenging to estimate the consequence of failure, as a variety of consequences can be identified such as safety, social, financial, and environmental.

A simplified alternative solution to numerically estimate risk score would be using a guideline, which predefines several levels for a chance of failure as well as consequences. For instance, a very low to a very high level of impact is classified by the organization for each type of consequence to help decision-makers to assign a level to a consequence of failure for each asset/system.

## 5.2.2 *Prioritization*

Prioritization analyses look into the cost-effective preservation and rehabilitation strategies based on life cycle costs. The most prevalent methods for prioritization are the benefit/cost ratio and the cost/effectiveness method. The output of prioritization is a list of projects requiring action, along with the timing and cost. Compared to ranking, in prioritization, the effectiveness of each treatment can be also tracked and several feasible treatments can be considered at a single (or multiple) point of time while more factors can be used in the decision-making.

These are the steps involved in a (multi-year) prioritization:

- Step 1. Forecast future condition (levels of damage indicator)
- Step 2. Treatment and timing options
- Step 3. Evaluate strategy effectiveness
- Step 4. Perform economic analyses
- Step 5. Use objective measures to prioritize needs
- Step 6. Project future needs

### 5.2.2.1 Forecast Future Condition (Levels of Damage Indicator)

To explain the steps, we will continue with the same example (Example 5.2). First, we forecast future conditions for 10 years using 3 deterioration curves: one for rutting, another for the cracking area, and the last one for the degradation of the structural capacity (measured by DBA). These curves are applicable to assets at different points of their lifespan, so for instance, Main St. is at 8 mm in terms of rutting, which is then expected to move to 11.5 after one year (Fig. 5.2).

The developed equations from the points are called deterioration models. Figures 5.3, 5.4, and 5.5 show developed models for a group of roads that belong to a given HG. The best-fitted curve is driven by ordinary least squares regression using the average values observed (See Sect. 3.2).

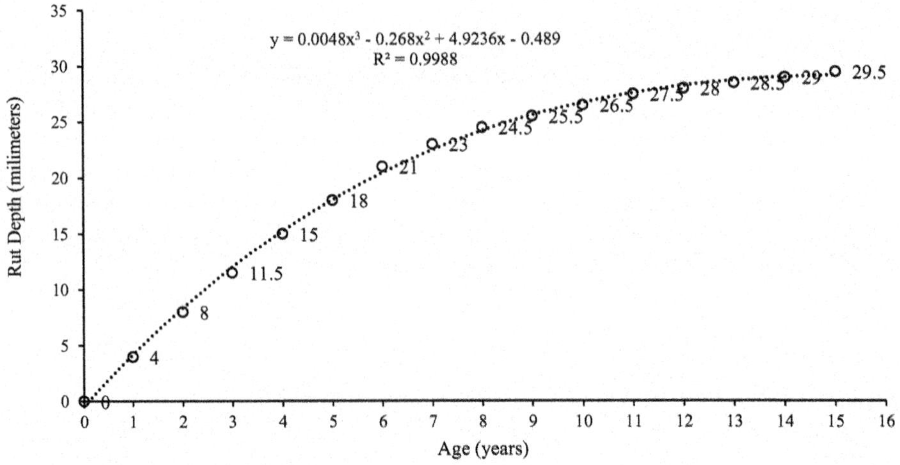


**Fig. 5.2** Sample rutting deterioration curve

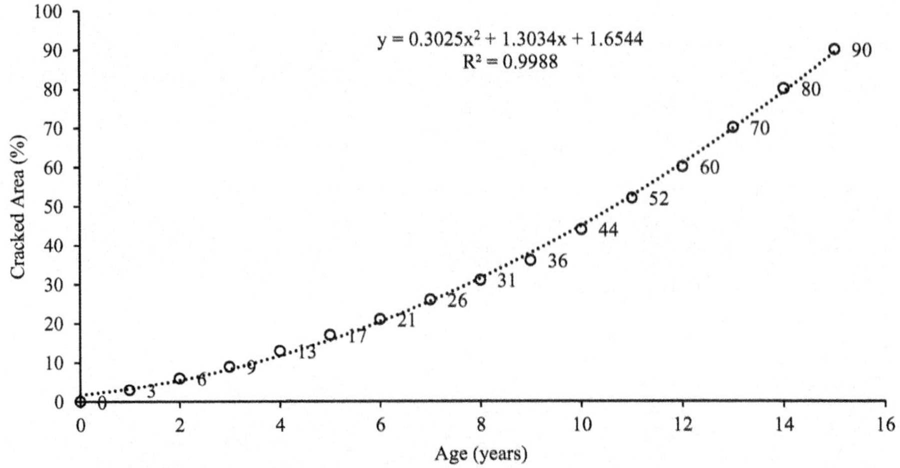


**Fig. 5.3** Sample cracking deterioration curve

#### 5.2.2.2 Treatment and Timing Options

We also have the decision rules that aid identify the best timing for treatment as shown in Table 5.10. They summarize the treatments available and their effectiveness. They are employed during the economic analysis.

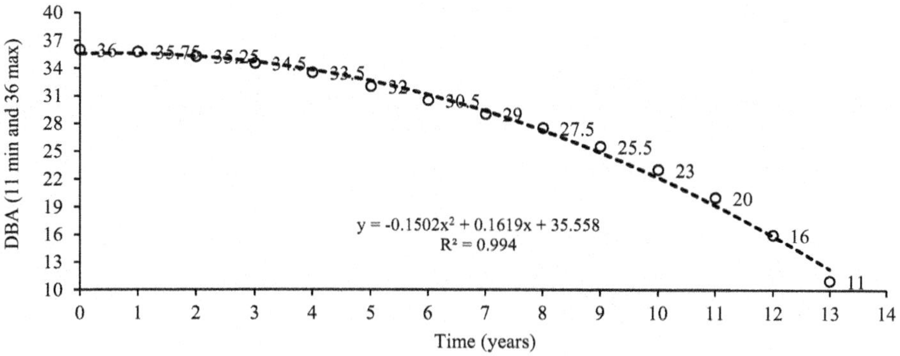


**Fig. 5.4** Sample DBA (structural capacity) decay curve

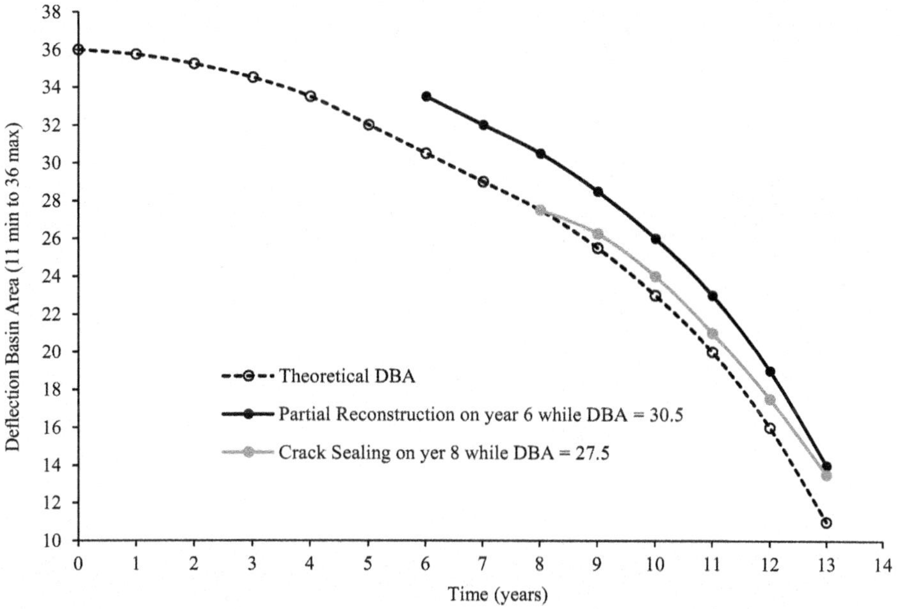


**Fig. 5.5** Decay and treatment effectiveness for crack sealing and partial reconstruction

**Table 5.10** Treatments and triggers

| Rules for treatments | Rutting | Cracking | DBA |
|---|---|---|---|
| Crack sealing | Less 3 | Less 10% | N.A. |
| Micro-surfacing | From 3 to 13 | N.A. | N.A. |
| Mill and overlay | More 13 | Less 20 | N.A. |
| Partial reconstruction | More 13 | More 20 | More 18 |
| Full reconstruction | More 13 | More 20 | Less 18 |

Notes: *N.A.* not applicable

#### 5.2.2.3 Evaluate Strategy Effectiveness

The rules of rejuvenation used in the previous examples are summarized in Table 5.11. They are estimated for data analyses of previous sections treated and the post-treatment trends.

Using the decision rules and taking advantage of the deterioration curves, we can consider the effectiveness of each treatment on the lifespan of the pavement. For example, consider the effectiveness of the structural capacity of partial reconstruction (applied to a pavement that exhibited a DBA of 30.5) and crack sealing applied to the pavement with 27.5 of DBA (Fig. 5.6).

**Table 5.11** Strategy effectiveness

| Treatment | Rutting effect | Cracking effect | Deflection basin area effect | Rate of deterioration of structure |
|---|---|---|---|---|
| Crack sealing | None | Reset to zero | None | 50% reduction[a] |
| Microsurfacing (including coat seal) | Reset to zero | Reset to zero | None | 50% reduction[a] |
| Mill and overlay | Reset to zero | Reset to zero | + 1 unit on DBA[a] | N.A. |
| Partial reconstruction | Reset to zero | Reset to zero | + 3 unit on DBA[a] | N.A. |
| Full reconstruction | Reset to zero | Reset to zero | Max. population value 34 (reset) | N.A. |

Notes: *N.A.* not applicable
[a]To be estimated from data

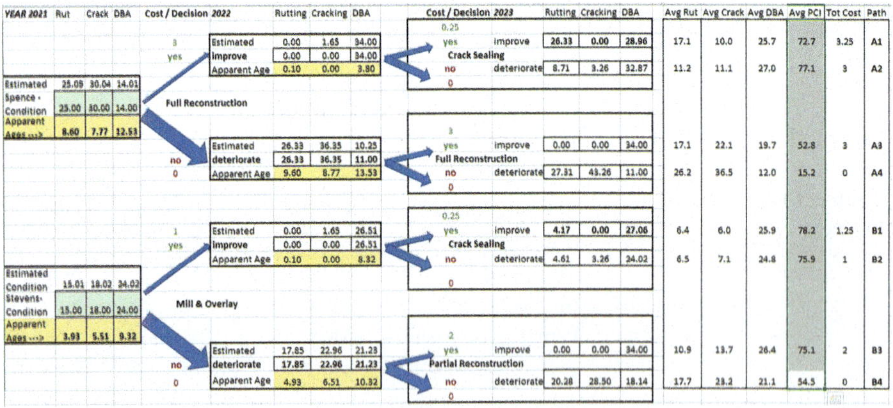


**Fig. 5.6** Sample decision tree for prioritization analysis of 2 segments in 2 years

#### 5.2.2.4 Perform Economic Analysis

Considering now the segments for Spence Avenue and Stevens Street only for the prioritization analysis along with the corresponding treatment as per the previous decision rules and the associated cost. As seen in Fig. 5.6, this opens eight different time paths, four for each road segment. But it does not stop there; there are 16 combinations possible for the four paths of the first segment and the four paths of the second segment. This is often called the decision tree from which economic analysis needs to be conducted to select the prioritized list of choices.

Let us now look at the combinations from both segments (16 in total) on the left portion of Table 5.12. In addition, on the right-hand side, we have the dominance analysis that eliminates 10 options that are inferior (or dominated by the combination given in the "Inferior" column).

So, if your budget is 4.25 million, you can choose combination A2 B1 (Spence full reconstruction 2022 and nothing on 2023, Stevens mill and overlay in 2022, and crack sealing in 2023); but if your budget is 4 million, your combination would be A2B2 (Spence full reconstruction 2022 and nothing on 2023, Stevens mill and overlay in 2022 and nothing in 2023); and so on. Budget levels are possible for 1, 1.25, 3, 4, and 4.25 million. Evidently, this oversimplified example ignores the reality of having to plan for several years and for networks with hundreds of segments.

**Table 5.12** Time path combinations and dominance analysis

| Combination | Total PCI | Total cost | Avg PCI |
|---|---|---|---|
| A1 B1 | 151 | 4.50 | 75.5 |
| A1 B2 | 149 | 4.25 | 74.3 |
| A1 B3 | 148 | 5.25 | 73.9 |
| A1 B4 | 127 | 3.25 | 63.6 |
| A2 B1 | 155 | 4.25 | 77.7 |
| A2 B2 | 153 | 4.00 | 76.5 |
| A2 B3 | 152 | 5.00 | 76.1 |
| A2 B4 | 132 | 3.00 | 65.8 |
| A3 B1 | 131 | 4.25 | 65.5 |
| A3 B2 | 129 | 4.00 | 64.3 |
| A3 B3 | 128 | 5.00 | 63.9 |
| A3 B4 | 107 | 3.00 | 53.6 |
| A4 B1 | 93 | 1.25 | 46.7 |
| A4 B2 | 91 | 1.00 | 45.5 |
| A4 B3 | 90 | 2.00 | 45.1 |
| A4 B4 | 70 | 0.00 | 34.8 |

| Combination | Total cost | Avg PCI | Inferior |
|---|---|---|---|
| **A2 B1** | **4.25** | **78** | **No** |
| **A2 B2** | **4** | **77** | **No** |
| A2 B3 | 5 | 76 | Yes A2B2 |
| A1 B1 | 4.5 | 75 | Yes A2B2 |
| A1 B2 | 4.25 | 74 | Yes A2B1 |
| A1 B3 | 5.25 | 74 | Yes A2B2 |
| **A2 B4** | **3** | **66** | **No** |
| A3 B1 | 4.25 | 66 | Yes A2B4 |
| A3 B2 | 4 | 64 | Yes A2B4 |
| A3 B3 | 5 | 64 | Yes A2B4 |
| A1 B4 | 3.25 | 64 | Yes A2B4 |
| A3 B4 | 3 | 54 | Yes A2B4 |
| **A4 B1** | **1.25** | **47** | **No** |
| **A4 B2** | **1** | **46** | **No** |
| A4 B3 | 2 | 45 | Yes A4B2 |
| **A4 B4** | **0** | **35** | **No** |

Imagine what this would look like for a network with a few hundred assets, a couple of decades, and more than one criterion (for instance, adding travel time, road safety, or any other criteria that are forecasted on time and can rejuvenate or change according to treatments scheduled). For this reason, mathematical optimization became common to solve this problem.

#### 5.2.2.5 Performance Targets and Funding Needs

The projection of future needs is a rather difficult task that requires the estimation of funding needs to accomplish desired performance targets (Fig. 5.7). For instance, consider in the example presented before (Table 5.12); we want to ensure certain performance goals in the form of pavement surface condition and structural preservation such as the rutting in the network not surpassing 14 mm and DBA to remain above 25. In such a case, the only feasible path is A2 B1 for a budget requirement of 4 million in the year 2022 and 25,000 in the year 2023.

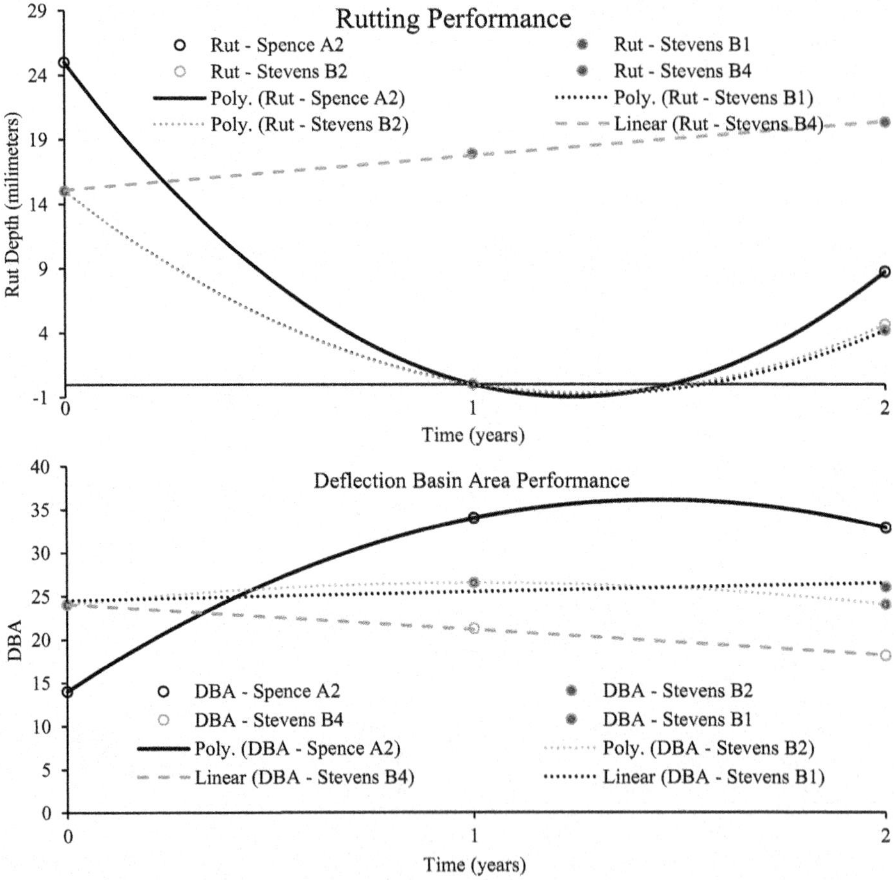


**Fig. 5.7** Performance projection and targets

#### 5.2.2.6 Use Objective Measures to Prioritize Needs

In the previous example, we utilized pavement condition and cost as the measures to prioritize and select time paths for each asset across time. However, there are other measures that could have been used to prioritize, such as road safety indicators, travel time, accessibility levels, vehicle operating cost, environmental gas emissions, etc. Further information on how to use these measures can be seen in Chap. 6.

### *5.2.3 Optimization*

Prioritization techniques improve the worst-first and ranking method in several aspects; however, still some limitations can be identified:

- In real cases, we have many assets while different scenarios could be implemented for each one. Thus, it means many alternatives and scenarios that are challenging to handle manually.
- As it was discussed, developing a multi-year plan is critical. Thus, decision-making would be more and more complex.
- Solving multiple treatment strategies would be much more complex.
- Even multi-year prioritization models disregard the impact of decisions for long-term objectives and can only capture medium-term consequences.

Like all other scientific problems, the rationale behind using optimization tools is enabling us to find the optimal solution for complex models with unlimited alternatives (i.e., many variables, several objectives, and constraints). For infrastructure with many assets and maintenance alternatives, long-term planning for 20 or 30 years is complex decision-making and a computing-intensive process that requires optimization techniques for its formulation and its solution.

#### 5.2.3.1 Decision Variables

As the main base of the optimization model, variables link decision alternatives to the model. In IAM, typically assets are variables that may or may not be selected for intervention. Thus, the next year's decision-making problem for infrastructure with 100 assets includes 200 (100*2) variables while each asset can be given two variables of 1 and zero (Eq. 5.4).

$$X_{s,t} = \begin{cases} 1 & \text{if action is taken on asset } s \text{, in year } t \\ 0 & \text{if no action taken on asset } s \text{, in year } t \end{cases} \tag{5.4}$$

It should be clarified that the total number of alternatives is usually much more than the number of variables. For the same example, if we assume that the budget forces asset managers to pick only 10 from 100 assets, then the total number of alternatives only for the first year would be $1.7 \times 10^{13}$. This number goes up if more than one feasible action can be selected for a single asset at the same time. For this reason, computer-based mathematical optimization would be a feasible solution to solve such problems.

#### 5.2.3.2 Dynamic Link

The formulation of the path that asset follows across time requires a dynamic link to connect each year with the following one, similar to the recursive formulations used in economics. Consider the asset "i" with a condition "Qi" on any given period of time "$t$" and the previous period "$t-1$" (often in terms of years); a simple decision variable of "yes" = 1 and no = 0 is used to represent the decision to provide the maintenance selected as per the decision rules (i.e., Table 5.12). If the maintenance is deployed, the asset will improve according to the effectiveness of the treatment "E", but if the decision is made to do not provide maintenance, the asset will deteriorate "D" (Fig. 5.8). This time dynamic is captured by the equation shown in Fig. 5.9.

If the maintenance happens on any given year "$t-1$", the variable $x_{tij}$ takes on the value of 1 and only the first term of the equation survives $(Q_{t-1} + E_j)$ and reports back a value of condition "$Q_t$". If the decision is made NOT to do the maintenance, then only the second portion of the equation survives and reports back a value "$Q_t$". This equation is the foundation for the formulation of the path of each asset for the entire analysis period, progressing forward from the current year through time until the last analysis year, opening branches of the decision tree, such as the one illustrated in Fig. 5.7, but for much more than 2 years.

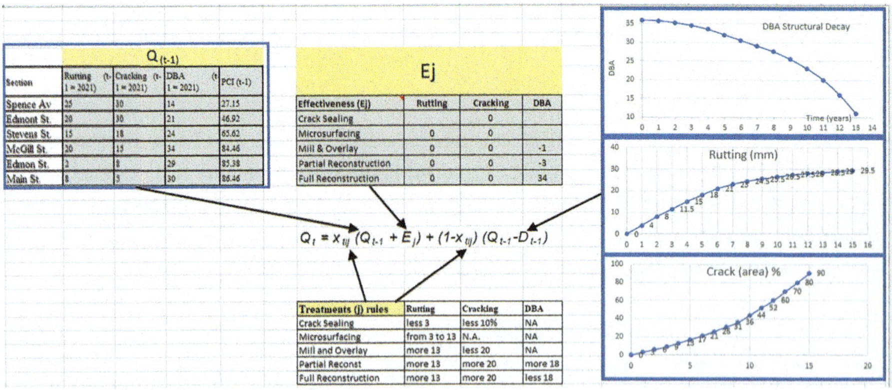


**Fig. 5.8** Dynamic link and its connection to the effectiveness, deterioration, and decision variables

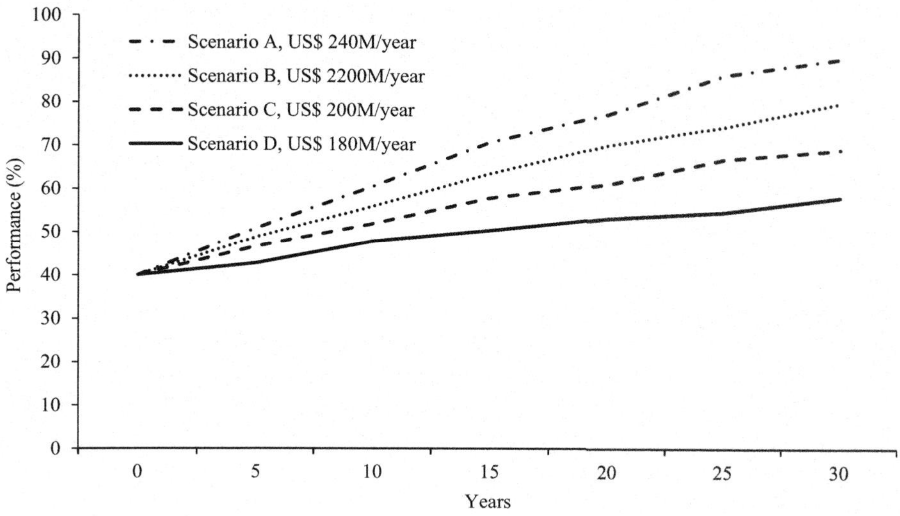


**Fig. 5.9** A sample long-term plan

#### 5.2.3.3 Objective Function

The next element necessary for any optimization is the objective equation. Simply put, the objective function establishes an expression that uses the decision variable to obtain either the minimum or maximum network-level aggregated values for a given key performance indicator.

Consider for instance the decision to maximize the level of DBA condition for the entire network or minimize the level of rutting or cracking for the previous example. Such objectives require a summation function and of the decision variable applicable for each asset, segment, or component candidate to receive maintenance, rehabilitation, upgrade, or reconstruction.

Let us continue with the previous example of road pavement sections, and write down the expression to maximize the DBA condition.

$$\max \sum_{i=1}^{n} \mathrm{DBA}_{tij} = \max \sum_{i=1}^{n} x_{tij} \left(\mathrm{DBA}_{t-1} + E_j\right) + \left(1 - x_{tij}\right) \times \left(\mathrm{DBA}_{t-1} - D_{t-1}\right) \quad \forall t \tag{5.5}$$

However, the previous expression could lead to an annual optimization for every year $t$. Instead, we need to attempt the maximum value for the whole analysis period as well, therefore expanding the expression as follows.

$$\max \sum_{t=1}^{T} \sum_{i=1}^{n} \sum_{j=1}^{J} x_{tij} \left(\mathrm{DBA}_{t-1} + E_j\right) + \left(1 - x_{tij}\right)\left(\mathrm{DBA}_{t-1} - D_{t-1}\right) \quad \forall t \tag{5.6}$$

Extending the previous equation to other indicators requires a simple replacement only. Of more interest will be to minimize the total cost for the whole network. This is based on the unitary cost ($C_{tij}$) on year "*t*" of treatment "*j*" for asset "*i*"; and the size ($S_{tij}$) on time "*t*" of asset "*i*" receiving treatment "j", which will be written for an annual basis as follows:

$$\min \sum_{t=1}^{n} x_{tij}\ C_{tij} S_{tij} \quad \forall t \tag{5.7}$$

To simplify formulation, a common assumption is considering a single treatment option for each condition (or age) range during asset lifecycle. For instance, as Fig. 5.9 shows, crack sealing will be implemented only when rutting is less than 3. Therefore, there would be only one treatment ($j$) at the same time, which means this J sigma can be removed from the formulation. This assumption is in line with best practices approaches and reflects reality.

Decision-making for IAM could be multi-objective, and different methods of goal programming or weighted average have been used by scholars to run optimization models. Chapter 6 provides some examples of multi-objective optimization in IAM planning.

#### 5.2.3.4 Constraints

The last piece of the puzzle in the formulation of an optimization problem is given by the constraints. They impose limits on the feasibility of the decision variable. For instance, an annual budget of USD 5,000,000 will be formulated as follows:

$$\sum_{t=1}^{n} x_{tij}\ C_{tij} S_{tij} < 5,000,000 \quad \forall t \tag{5.8}$$

Another common constraint is the desire to accomplish a minimum or maximum level of a given indicator; for example, average network rutting for time $t$ ($\text{RUT}_{tij}$) to be less than 14 mm would be formulated as follows:

$$\frac{\sum_{i=1}^{n} \text{RUT}_{tij}}{\sum_{i=1}^{n} S_{tij}} < 14, \text{ where } \text{RUT}_{tij} = x_{tij}\left(\text{RUT}_{t-1} + E_j\right) + \left(1 - x_{tij}\right)\left(\text{RUT}_{t-1} - D_{t-1}\right) \tag{5.9}$$

#### 5.2.3.5 How to Solve an Optimization Problem

There are two major techniques to solve an optimization problem: (1) Heuristic methods and (2) linear (non-linear) programming. Heuristic methods use an engine to generate values of the decision variable (either 1 or zero) for all possible branches across the decision tree formulated as in Sect. 5.2.2.4. There are multiple engines to do so, being the most simple, the one called genetic algorithm consists of a random binary number generator, a record of the sequence (chain) of binary numbers, and the corresponding value of the objective function and the constraints while eliminating any iteration that produces a violation of the constraints and saving those that do not.

One of the main problems with this approach is that there is no way to ensure one has accomplished the optimal value; always the analyst will end up with a suboptimal value whose quality will depend on the number of iterations and the ability to identify specific binary sequences that produce superior values and implementing some mechanism to control the evolution by keeping those portions of the binary sequence that support the accomplishment of the objective (maximization or minimization) and generating new values for those portions which do not. This approach is called heuristic evolutionary, with specific tactics such as simulated annealing, ant colony, and many others, which the reader can find in optimization books.

For what matters in this book, the heuristic approach is a fast way to obtain results, but the only way to accomplish the true optimal value is by utilizing linear programming. Linear programming uses the simple algorithm in its foundation and visits the feasible space of the problem as delineated by the constraints trying to move in the direction of the objective (maximize or minimize) until the last point of the feasible space is found and such point delivers the binary sequence of the decision variable.

Unfortunately, running linear programming with time dependencies (dynamic) and for a binary variable for thousands of assets and multiple years is computationally expensive and poses a burden on the capability of the computer being used.

For these reasons, the best approach to handle the optimization of infrastructure networks and facilities is to prepare the optimization problem and define the mechanisms for the time dependencies with the dynamic link, the treatment rules, the cost, and the effectiveness of the treatments in the first instance in any desired platform, and handle the solution via commercial solvers (such as MOSEK [Mosek ApS, 2020], LINDO [Lindo System, 2020], CPLEX [IBM, 2020], etc.) or non-commercial (such as Excel solver), which will reduce the problem by eliminating the inferior paths, and those that violate the constraints and solve the problem through advanced matrix algebra techniques, which is out of the scope of this book.

The ideal approach is to run the optimization model for the entire time window (e.g., 20 years). For example, using Eq. 5.6 instead of 5.5 or minimizing the total cost of the whole period of time. However, running an optimization model, in this case, would be more challenging and may need to use commercial solvers and language programming. This should not stop the team to take advantage of optimization and long-term planning. The alternative simplified solution, which can be handled even via non-commercial (such as Excel solver), would be running optimization year by year while each year's decisions depend on the previous year's actions. Still, this approach guides the team toward achieving objectives, however, may not lead to the best scenario. For instance, the final solution to minimize (as an objective) the cost of each year in 20 years plan (running optimization 20 times) can be close but is not necessarily the same as running one optimization to minimize the total costs for the whole 20 years.

#### 5.2.3.6 Optimization Scenarios

The mathematical formulation in optimization models can be defined differently; however, the key factor in selecting the right scenario is respecting organization requirements, constraints, goals, and policy (Amador-Jimenez & Mohammadi, 2020). In the context of IAM, these are the most common scenarios:

**Scenario A** How much budget do I need to maintain or achieve a non-declining overall condition (Q)?

$$\textbf{Minimize: } Z = \sum_{t=1}^{T}\sum_{i=1}^{I} C_{t,i}.X_{t,i}.L_i \longrightarrow Z_t = \sum_{i=1}^{I} C_{t,i}.X_{t,i}.L_i$$

$$\textbf{Subject to} \quad \sum_{i=1}^{I} Q_{t,i}.L_i \geq \sum_{i=1}^{I} Q_{t-1,i}.L_i$$

$$Q_{t,i} = X_{t,i}(Q_{t-1,i} + q_{t,i}) + (1 - X_{t,i})(Q_{t-1,i} - d_{t,i})$$

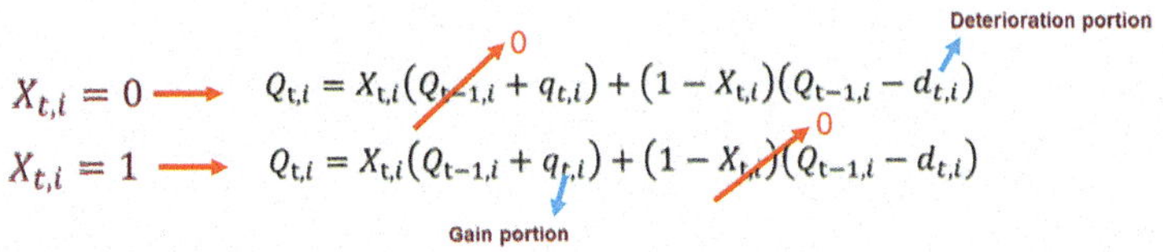


where

$Q_{t,i}$ = Overall condition at year $t$, of asset $i$ after maintenance or do nothing
$x_{t,i}$ yes = 1 and no = 0
$L_i$ = size of the asset
$C_{t,i}$ = Unit cost of treatment on asset $i$ at year $t$.
$B_t$ = budget for year $t$
$q_{t,I}$ = Improvement portion of condition from the year $(t - 1)$ to year "$t$" for asset "$i$"
$d_{t,I}$ = Dropped portion of condition from the year $(t - 1)$ to year "$t$" for non-selected asset "$i$"

**Scenario B** How much budget do I need to achieve a target overall condition ($\mathrm{LOS}_n$)?

$$\textbf{Minimize: } Z_t = \sum_{i=1}^{I} C_{t,i}.X_{t,i}.L_i$$

$$\sum_{i=1}^{I} Q_{t,i}.L_i \geq (LOS_n) \sum_{i=1}^{I} L_i$$

Term is fixed, constant

$$Q_{t,i} = X_{t,i}(Q_{t-1,i} + q_{t,i}) + (1 - X_{t,i})(Q_{t-1,i} - d_{t,i})$$

And $\mathrm{LOS}_n$ = Target level of service for the network, $n$ (when you have more than one network of assets and each has its own target).

**Scenario C** Given that I have a fixed budget per year ($B_t$), what is the best overall condition achievable?

$$MAXIMIZE\!: Z_t = \sum_{i=1}^{I} Q_{t,i}.L_i \quad \textbf{OR} \quad MAXIMIZE\!: Z_t = \frac{\sum_{i=1}^{I} Q_{t,i}.L_i}{\sum_{i=1}^{I} L_i}$$

$$0 < \sum_{i=1}^{I} C_{t,i}.X_{t,i}.L_i \leq B_t$$

$$Q_{t,i} = X_{t,i}(Q_{t-1,i} + q_{t,i}) + (1 - X_{t,i})(Q_{t-1,i} - d_{t,i})$$

## 5.3 Asset Management Plan

By completing decision-making, asset managers would be able to extract planning outputs. Asset management plan (AMP) is a typical key report for organizations, which presents the AM path. This document is a tool to define, track, and manage SGR projects communicating internally with various teams including engineering, finance, operation, project delivery, and senior managers. AMP should be an alive and comprehensive report, which includes but is not limited to these sections:

### *5.3.1 Goals and System KPIs*

What are the main KPIs and targets to achieve organizational goals? What are the current standing points and what would be the goals for the future?

### 5.3.2 *Assets Portfolio*

This section provides a summary of registered assets including the location, quantity, current condition/performance, and RV.

### 5.3.3 *Plans*

Different types and layers of plans such as long-term (strategic), medium-term (tactical), and short-term (operational) can be provided in this report. AMP can be updated yearly or longer such as every 3 years base and the plan will be presented accordingly. It is supposed to provide an operational plan including a list of assigned interventions/budgets for selected assets/systems in the coming year/years. At the same time, tactical and strategic plans provide a longer-term view of future investments and achievements respecting KPIs. Figure 5.9 presents a sample long-term plan analyzing different budget scenarios. LOS in this figure can be any type of KPIs (e.g., PCI for pavement, BCI for bridge, and FCI for building), while the impacts of different budget scenarios are investigated. These kinds of analyses are key tools for policymakers to capture the future impacts of current decisions and to select the appropriate approaches addressing all stakeholders' concerns and expectations.

## 5.4 Exercises

**Exercise 5.1**
A recent assessment for a city is partially presented in the below table. If you are assigned \$1,500,000 for the first coming year, how do you plan for maintenance?

A. If you don't care about traffic
B. Considering traffic as a weighting factor

**Note:** Construction team recommended to do minor maintenance for $60 \leq \text{PCI} < 90$ (\$10,000 per kilometer per lane), major maintenance for $40 \leq \text{PCI} < 60$ (\$50,000 per kilometer per lane) and reconstruction for $\text{PCI} < 40$ (\$200,000 per kilometer per lane) (Table 5.13).

**Exercise 5.2**
A road agency recently built a highway, and experts suggested three different scenarios for maintenance. The AADT is predicted to be 85,000, and inflation could be neglected. The cost for minor maintenance, major maintenance, and reconstruction are \$10,000, \$50,000, and \$200,000 per lane per kilometer, respectively. The interest rate is 5% annually. Scenario 1 includes doing a minor after every

**Table 5.13** Exercise 5.1 dataset

| Segment | AADT | Current PCI | Length (km) | Number of lanes |
|---|---|---|---|---|
| A | 353 | 66 | 6.5 | 2 |
| B | 450 | 88 | 5 | 2 |
| C | 7000 | 45 | 3 | 3 |
| D | 3500 | 50 | 3 | 3 |
| E | 21,000 | 69 | 2.5 | 4 |
| F | 15,500 | 52 | 3.25 | 4 |
| G | 9900 | 75 | 4 | 3 |
| H | 18,000 | 40 | 2.8 | 4 |
| I | 800 | 35 | 3.6 | 2 |
| J | 6000 | 55 | 3.56 | 3 |

**Table 5.14** Exercise 5.3 dataset (1)

| Segment | Current condition (PCI) | Length | Deterioration rate |
|---|---|---|---|
| s1 | 75 | 5 km | 5 points/year |
| s2 | 50 | 10 km | 5 points/year |

4 years four times. The second scenario would be doing minor after 4 years and then doing major after 8 years from minor maintenance. Finally, doing reconstruction after 16 years. Answer the below question briefly and to the point:

- Which scenario has a minimum life cycle cost ($) (only cost related to maintenance and construction)?
- Which scenario does not look to be a feasible approach? Why?
- Which scenario may provide the highest service for users (ignore feasibility)? Why?
- Which scenario do you pick? Why?

**Exercise 5.3**

Develop a two-year decision tree for a road network of two segments and answer the below questions:

- What would be the required budget to achieve the highest network PCI?
- What would be the network PCI for a no-budget scenario?
- What would be the best possible achievement (i.e., network PCI) for a total available budget of $ 2,500,000? (Tables 5.14 and 5.15)

**Table 5.15** Exercise 5.3 dataset (2)

| Treatment | Operation window | Cost per km | Effectiveness |
|---|---|---|---|
| 1—Crack sealing | 70–90 | 25,000 | Gain 5 points |
| 2—Micro-surfacing | 60–69 | 50,000 | Gain 30 points |
| 3—Thin overlay | 50–60 | 200,000 | Gain 40 points |
| 4—Reconstruction | 0–50 | 500,000 | Brand new |
| 5—Do nothing | 91–100 | 0 | None, deteriorates |

## References

Amador-Jimenez, L., & Mohammadi, A. (2020). Decision making methods to prioritise asset-management plans for municipal infrastructure. *Infrastructure Asset Management*, 1–14.

IBM. (2020). CPLEX software. https://www.ibm.com/

Lindo System. (2020). LINDO software. https://www.lindo.com/

Mohammadi, A., Igwe, C., Amador-Jimenez, L., & Nasiri, F. (2020). Applying lean construction principles in road maintenance planning and scheduling. *International Journal of Construction Management*, 1–11.

Mosek ApS. (2020). MOSEK software. https://www.mosek.com/

# Chapter 6
# Emerging Issues and New Expectations

## 6.1 Introduction

In this chapter, novel ideas in IAM to enhance the classical platforms, as well as to address newly raised concerns, will be presented. The chapter explains how asset managers dealing with infrastructure such as transit systems can improve the performance of the system and enhance the efficiency of current practices as well as look at other objectives, which are not typically under attention as AM goals. Several sample models recently published in scientific journals are explored.

To date, most IAM systems revolve around the physical condition that does not necessarily represent the best interest of the users. For instance, the IRI as representative of pavement surface condition fails to indicate the safety of road users (pedestrians, cyclist, motorist) or the degree of accessibility (ability to reach socio-economic opportunities) or mobility (ability to move in a timely manner) from one location to another (for example, from home to work at 7 am). Another example can be the level of comfort (e.g. vibration, thermal, and noise) in the railway cars while maintenance planning mostly focuses on the reliability and condition of assets, which may or may not address customer concerns for comfort (Mohammadi et al., 2019). In fact, classical IAM efforts are often asset based or condition based rather than user based or service based.

It is also important to notice that there are aspects of infrastructure management that fundamentally change when abnormal circumstances arrive, such as in the event of natural hazards (coastal or inland flooding, earthquakes, ice storms), and as climate change seems to increase temperatures and result in more frequent and more severe storms, civil infrastructures experience the need for retrofitting to cope with the new circumstances.

Attention was paid to environmental issues and their relation to infrastructure management a decade ago, in the early 2010s. The main aspect incorporated within infrastructure management was that of pollution provoked by maintenance,



A. Mohammadi, L. Amador Jimenez, *Asset Management Decision-Making For Infrastructure Systems*, https://doi.org/10.1007/978-3-030-97614-9_6

rehabilitation, upgrade, and expansion of current infrastructure projects (Faghih-Imani & Amador-Jimenez, 2013).

Finally, a restricted budget as the most common barrier to taking care of infrastructure brings a need for a smart investment and avoiding non-value-adding activities.

## 6.2 Advanced IAM Platforms

Therefore, future IAM is expected to incorporate our current and future concerns. In the coming sections, examples are provided from novel methodologies proposed in the literature to address the below main global and local concerns.

- Environmental concerns such as GHG emissions
- Sustainability concerns such as energy efficiency
- Human development
- Level of comfort
- Climate change
- Travel time
- Vulnerability and resiliency issues

### *6.2.1 Environmental Concerns in Pavement Management*

Faghih-Imani and Amador (2013) proposed a model to incorporate GHG emissions and energy usage in the PMS. Several treatment alternatives can be identified for pavement interventions that do not necessarily have the same impacts on the environment. Maintenance activities for pavement usually consume high energy, producing $CO_2$, which is the main factor in GHG emissions. The model proposed a performance-based optimization platform while it measured the environmental footprint of each type of treatment incorporated in the decision-making process. The objective was to find the best scenario for pavement interventions while seeking for minimum required budget, maximizing performance, and minimizing energy usage and GHG emissions. To achieve these objectives, three trade-off steps were taken:

Step I: Minimizing budget subject to the non-declining condition

In this step, the model tried first to identify the minimum required budget while the network did not lose the current (at the starting point) overall performance. Thus, the current overall condition (i.e., IRI) as a constraint and minimizing cost as an objective were applied in the mathematical formulation.

Step II: Maximizing condition subject to the available budget

This is the most common scenario in optimization platforms, which reflects the reality in IAM. The model is run using a predefined available budget by

**Table 6.1** Treatment Alternatives

| Treatment | Reconstruction | Major Rehab | Hot In-Place Recycling | Chip Seal | Micro-surfacing |
|---|---|---|---|---|---|
| Details | 100 mm HMA over 150 mm Aggregate Base | 100 mm Overlay | Thickness 5 cm 50/50 Recycle/new | Emulsion 2.0 L/m$^2$ Aggregate 21 kg/m$^2$ | Type III, 12% Emulsion, 13 kg/m$^2$ |
| Life Extension (years) | As New | 15 | 10-May | 6-Mar | 5-Mar |
| Energy Use Per Year (MJ/m$^2$) | 9.9 | 9.2 | 6.5–13 | 1.5–3 | 1.3–2.2 |
| GHG Emissions Per Year (kg/m$^2$) | 0.7 | 0.8 | 00.5–1.0 | 0.08–0.10 | 0.06–0.10 |

Adapted from Faghih-Imani and Amador (2013)

**Table 6.2** Defined Scenarios

| Scenario | A | B | C |
|---|---|---|---|
| Objective | Minimize Cost | Maximize Condition | Minimize Energy Use and GHG emission |
| Constraint | Non-increasing IRI | Annual Budget from A | Annual Budget from A and network's average IRI from B |
| Outcome | Annual Budget | Network's Average IRI | Sustainable choice of treatments |

Adapted from Faghih-Imani and Amador (2013)

policymakers for a given time (i.e., constraint) seeking to maximize the overall condition (i.e., objective).

Step III: Minimizing energy usage and GHG emissions subject to the acceptable condition (found in Step II) and budget. This step presents the novelty of this work.

$$\text{Minimizing } Z = \sum\sum\sum \text{Energy Usage} + \sum\sum\sum \text{GHGs Emissions}$$

$$\text{Subject to}: \sum\sum\sum \text{Total Cost} \leq \text{Available Budget}$$

$$\text{Subject to}: \sum\sum\sum \text{Overall Condition} \geq \text{Acceptable Condition}$$

This model was applied to a real case study, and Table 6.1 presents the treatment windows; while for each alternative besides its life-extension effectiveness, the produced GHG emissions (Kilogram/m$^2$) and consumed energy (Megajoule/ m2) are identified. Three scenarios (based on three steps) were defined (Table 6.2).

Figures 6.1, 6.2, and 6.3 present the results of this study. Figure 6.1 shows the impact of each scenario on the overall condition (IRI). Using the result of the first run (Scenario A), a constant budget per year was used as a constraint on the second

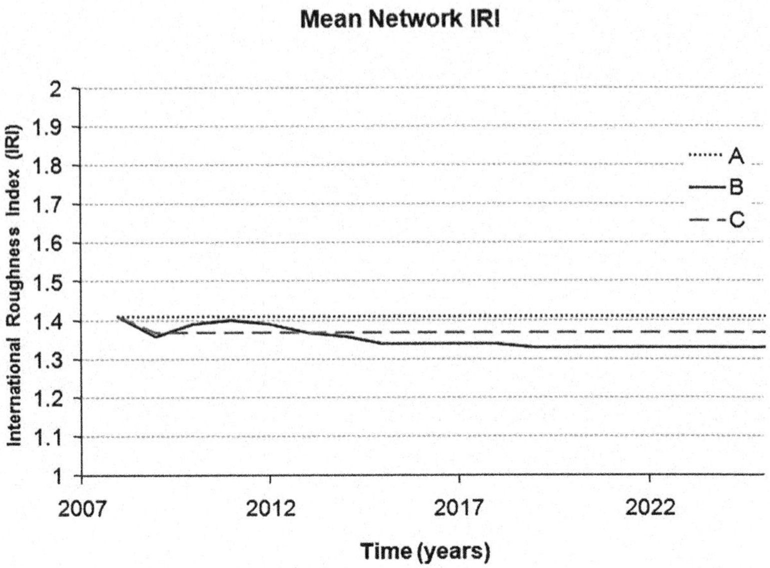


**Fig. 6.1** Comparing overall performance for Scenarios A, B, and C. (Faghih-Imani and Amador (2013))

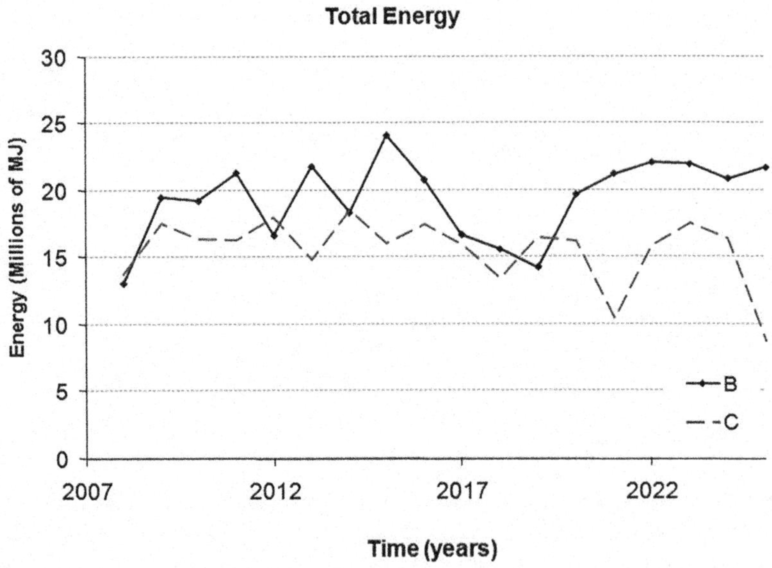


**Fig. 6.2** Comparing total energy usage for scenarios A and B. (Faghih-Imani and Amador (2013))

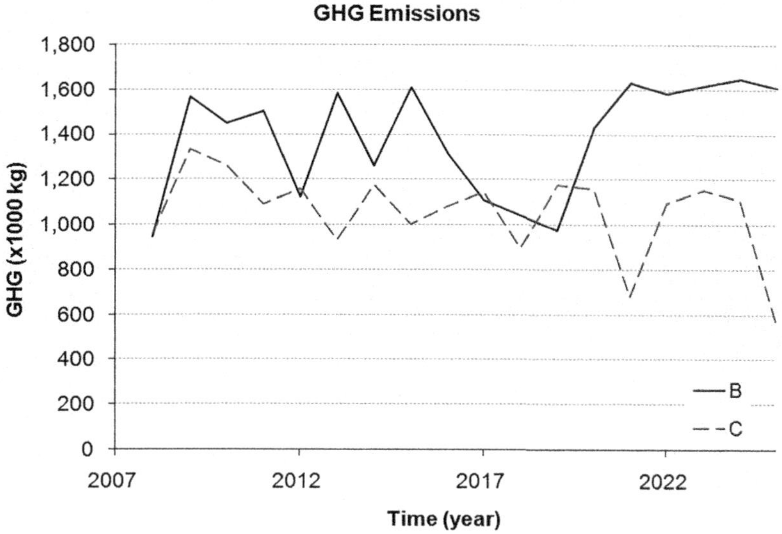


**Fig. 6.3** Comparing GHG emissions for scenarios A and B. (Faghih-Imani and Amador (2013))

analysis (Scenario B) to maximize the level of service (here, minimizing IRI). This scenario is usually the core of current pavement management systems. The result of this analysis was a cost-effective set of treatments to maximize the network's level of service using the planned annual budget. The last analysis incorporated the environmental impacts of each treatment. The goal of this scenario was to identify a set of treatments that could minimize the amount of GHG emissions and energy use while using the same budget and attaining almost the same condition as scenario B. Thus, scenario C was defined as minimizing energy use and GHG emissions of pavement maintenance and rehabilitation work subject to the same budget and almost the same network's average IRI as scenario B. Not considering the impacts of road users such as traffic delays and congestions in this scenario is a significant limitation of this study. As was expected, scenario A maintains an initial average of the network, while scenarios B and C were able to reach similar levels.

However, Figs. 6.2 and 6.3 indicate the efficiency of the model to reduce the negative impacts on the environment. As shown, the difference in condition is negligible while scenario C tries to reach minimum effects on the environment while spending a similar budget as scenario B. For more details, readers can refer to Faghih-Imani and Amador (2013).

### 6.2.2 Human Development

Transit systems play a critical role in achieving sustainable development goals (SDGs) in all nations, particularly under-developing societies (UN, 2016). Improving socio-economic factors such as employment, poverty alleviation, education, and gender equality in any society can be pushed by reliable, safe, and accessible transit systems. Classically, these factors are in the center of decision-making for expanding transit networks rather than transit asset management (TAM); however, Mohammadi et al. (2018) proposed a model to incorporate multiple human development criteria and sub-criteria in AM for the railway systems. Traditionally, the model is supposed to seek maximizing overall LOS in the network; however, the proposed methodology showed how transit agencies can improve typical KPIs enhancing human development concerns. Figure 6.4 presents the general idea behind this methodology.

A specific methodology was developed to measure quantitatively the level of HD for each alternative. Then, a multi-objective decision-making and optimization model was utilized through weighted formulation where α and β indicate the corresponding preference weights for both objectives of maximizing Overall LOS (OLOS) and Overall HD Index (OHDI).

$$\text{Maximizing } Z = \sum\sum\sum \text{Overall LOS} + \sum\sum\sum \text{Overall HDI}$$

$$\text{Subject to}: \sum\sum\sum \text{Total Cost} \leq \text{Available Budget}$$

This platform was also applied to a real railway case. The classical approach (only LOS) was compared with the proposed formulation in this study (combined-objective), and Figs. 6.5 and 6.6 present the results of this study. Fig.6.5 shows the predicted OLOS for this long-term planning for both scenarios.

Figure 6.6 presents the impact of each scenario on enhancing OHDI in the impacted regions by this railway system. The main point is that the novel methodology was able to push investment for poor neighborhoods and improve the OHDI while it did not lose OLOS, which could be a concern for combining these objectives. For more details, readers can refer to Mohammadi et al. (2018).

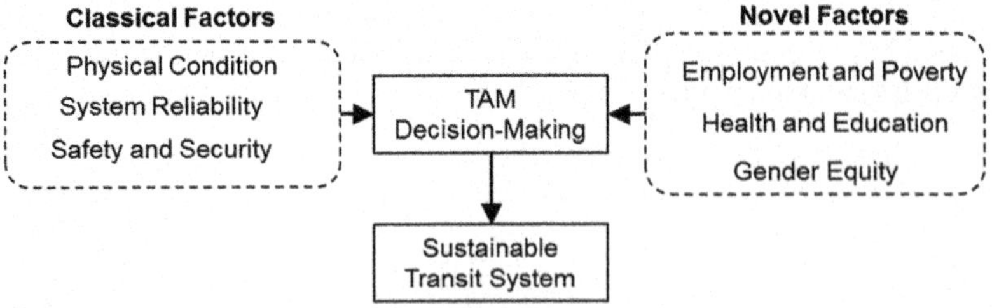


**Fig. 6.4** Classical and novel factors in TAM. (Adapted from Mohammadi et al. (2018))

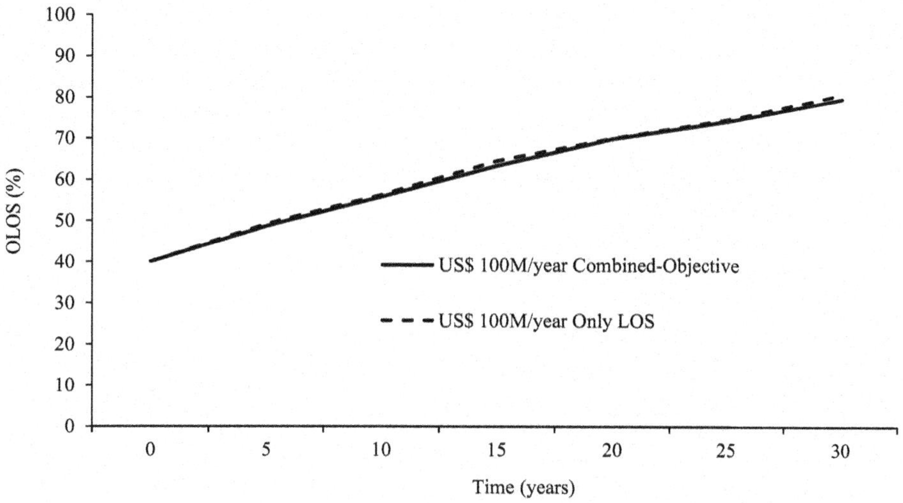


**Fig. 6.5** Comparison of OLOS in combined-objective and only LOS objective. (Mohammadi et al. (2018))

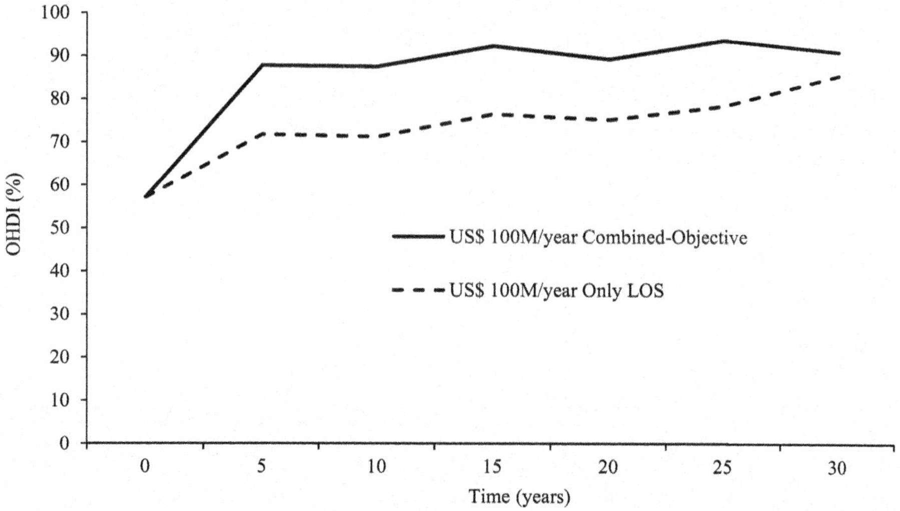


**Fig. 6.6** Comparison of OHDI in combined-objective and only LOS objective. (Mohammadi et al., (2018))

### *6.2.3 Level of Comfort*

A comfortable ambiance results in increased drivers' performance and improved users' safety. A greater degree of passengers' satisfaction and, subsequently, a higher likelihood of using public transport would be achievable by providing a better service quality in public transportation. Therefore, customers' expectations and concerns need to be monitored and analyzed by public transit agencies in order to be addressed later in IAM platforms as well as demand prediction. A primary step in this regard would be measuring quantitively the level of comfort using the key criteria. The main objective of the proposed study by Mohammadi et al. (2019) was to establish numerical assessment indices for railroad travelers' comfort and safety incorporating factors as diverse as humidity, temperature, vibration, the concentration of $CO_2$, noise, and lighting levels inside the vehicles based on a set of maximum and minimum thresholds identified with respect to public comfort and health standards. Next, the comfort indices of different factors can be combined through a global index. Such a global index could be used by decision-makers to objectively distribute budget for maintenance and rehabilitation actions, and it could help transit planners to capture reality in demand prediction modeling (i.e. transit mode choice analysis) and particularly useful for agencies when communicating with the public and the government (Fig. 6.7). As was discussed in Sect. 2.3, public infrastructure is mainly assessed from two main aspects of technical and social. The level of comfort addresses sociological concerns of performance, which are commonly ignored in the current practices.

Comfort indices could be used by asset managers to assess the entire network accordingly and objectively distribute budgets for maintenance, rehabilitation, and replacement actions leading to the improvement of comfort levels. Each comfort factor represents one or more assets condition and performance not only inside of the railcar but also in rail track and even tunnels. Figure 6.8 shows how the main comfort factors in urban railway journeys and their corresponding assets link railway AM to the level of comfort.

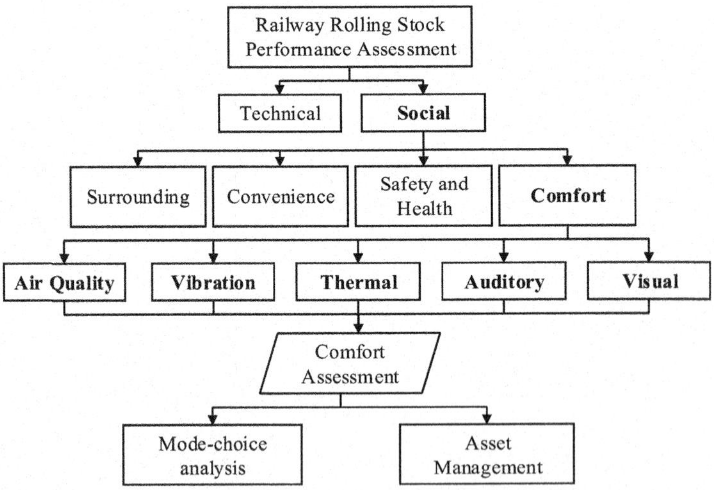


**Fig. 6.7** Level of comfort study road map. (Mohammadi et al. (2019))

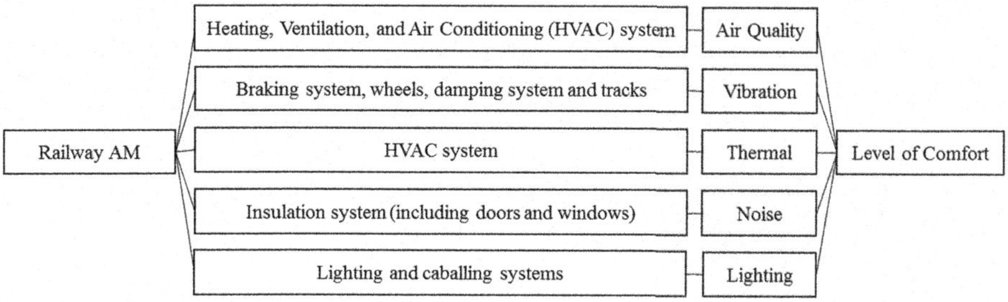


**Fig. 6.8** Level of comfort in railway cars and AM. (Adapted from Mohammadi et al. (2019))

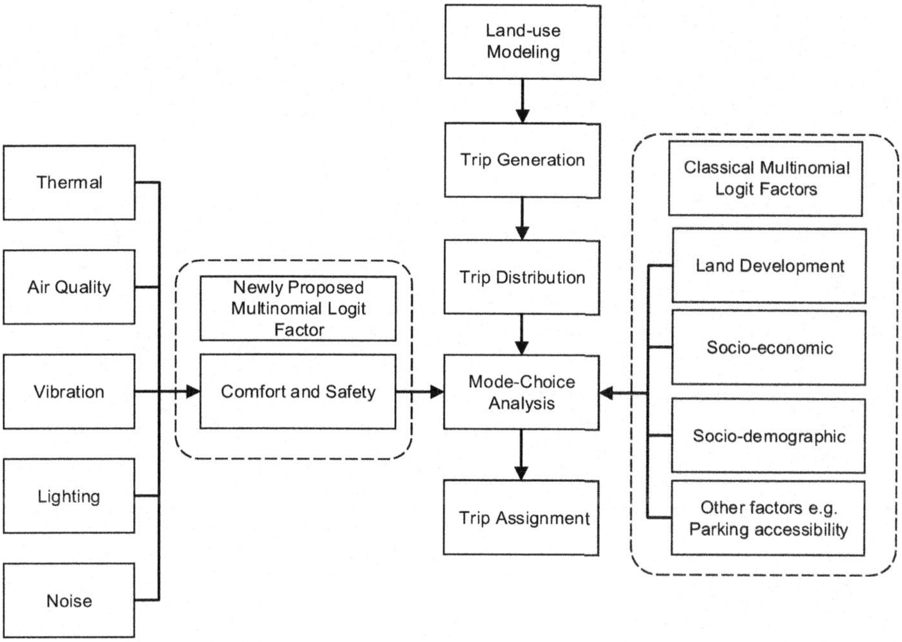


**Fig. 6.9** Addressing the level of comfort in mode-choice analysis and demand modeling. (Mohammadi et al. (2019))

Meanwhile, as a common approach for demand prediction, disaggregate discrete choice models such as four-step land-use modeling calculate a maximum likelihood for each travel choice based on a trip utility function, which is a weighted function comprised of travel time and cost, and other classical influencing factors. However, many best practices proved the role of the level of comfort in mode-choice. Thus, capturing comfort assessment can enhance demand prediction reflecting reality (Fig. 6.9). The idea can also be expanded to trip advisor apps in smartphones to provide an extra feature of comfort level for users (Amador et al. 2017).

For this purpose, data were collected and analyzed using low-cost mobile sensor technologies and smartphones in a real case network. The proposed model can handle traditional buses, bus rapid transit (BRT), light rail transit (LRT) , and subways to guide planning for their maintenance, upgrade, and expansion to achieve higher levels of convenience and reliability which then encourage higher transit ridership. For more details, readers can refer to Mohammadi et al. (2019).

### 6.2.4 Climate Change

Climate change is influencing our infrastructure more than before and probably less than in the future! Planning for any infrastructure must be linked to climate change issues, particularly for longer time strategies. For instance, the severity and frequency of storms have been changed gradually, and it is also expected to see higher rainfall intensity in the future. This directly impacts storm pipeline networks as a vital urban infrastructure causing several already reported flooding and is linked to the resiliency of the system. Thus, managing these assets can not be handled unless direct and indirect consequences of climate change are captured in medium- and long-term planning processes.

Traditionally, lucky municipalities replace a storm pipeline when it passes service life or lost performance. In such a situation, usually the old pipe will be replaced by a brand-new pipe of the same size. A bigger pipe might be used where a significant change in landscape is observed to avoid flooding. However, changes in rainfall intensity also can result in higher demand (required higher capacity of storm pipe). A smart IAM considers the impacts of climate change and captures its consequences in long-term decision-making. Amaodr et al. (2020) proposed a methodology for storm pipeline maintenance and replacement while the impact of changes in the landscape as well as climate change was addressed in AM process (Fig. 6.10).

The model defined a mathematical optimization formulation as below where the model tried to maximize overall pipe network conditions and minimize the demand over capacity ratio. This ratio represents the risk of flooding by comparing the network capacity and demand for water discharge during storms, while the future change in demand was also estimated.

$$\text{Maximizing } Z = \sum\sum\sum \text{Overall Condition} - \sum\sum\sum \text{Overall } \frac{\text{Demand}}{\text{Capacity}}$$

$$\text{Subject to : } \sum\sum\sum \text{Total Cost} \leq \text{Available Budget}$$

Thus, as a key element of long-term planning for the asset, the performance prediction model should be developed for condition as well as the demand-capacity ratio, which was discussed earlier in Sect. 3.5.1. Figure 6.11 shows developed demand-capacity prediction models (as the second performance indicator in this model) for three HGs.

The proposed model was implemented in a case study. Figure 6.12 shows the results of the running model for 25 years. As can be seen, the model was able to improve overall storm network performance (average condition) while dropping the

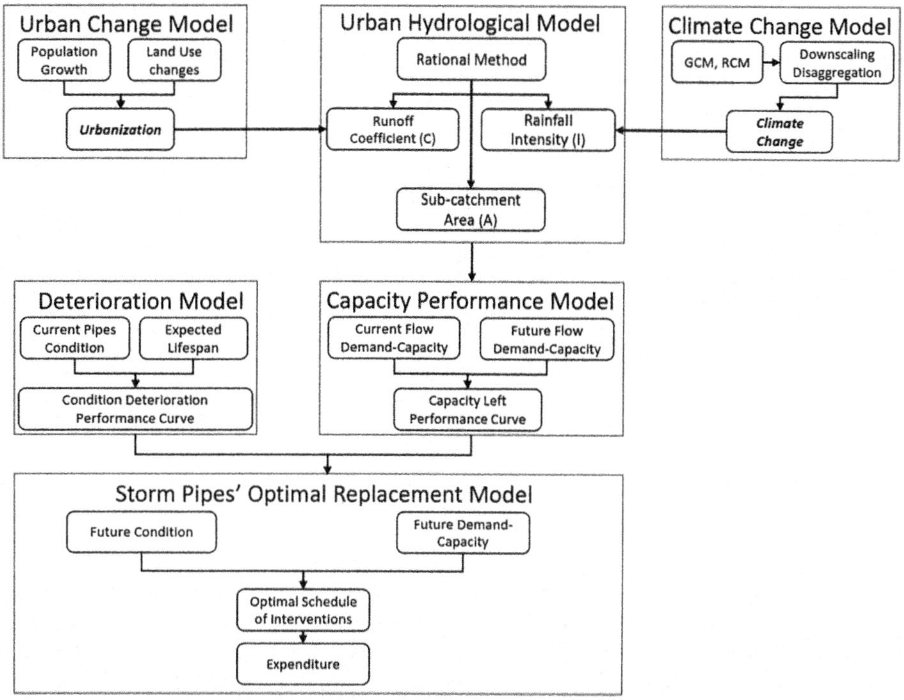


**Fig. 6.10** Research methodology for incorporating climate change in IA. (Amador et al. (2020))

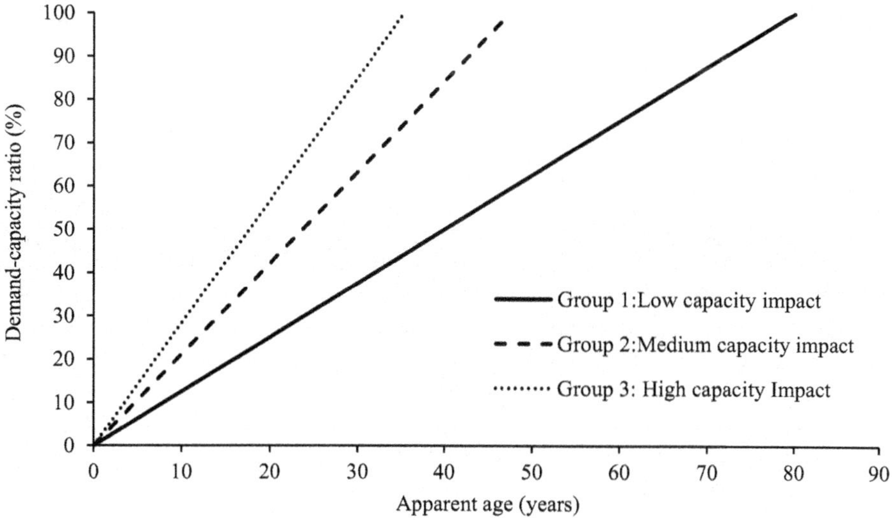


**Fig. 6.11** Demand-capacity prediction model. (Amador et al. (2020))

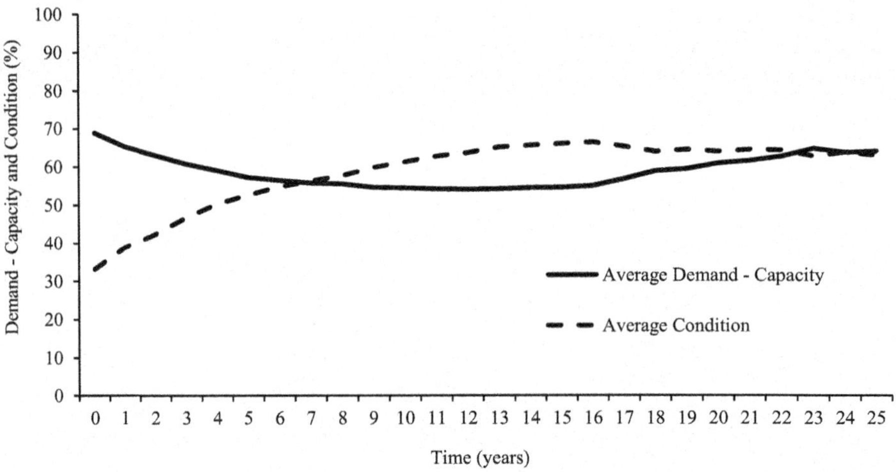


**Fig. 6.12** Long-term investment achievements for both objectives. (Amador et al. (2020))

demand over capacity ratio meaning that this methodology improved the resiliency of the network and decreased the risk of flooding.

Similar concerns can be addressed for other infrastructure. For instance, heavy storms, flooding, or changing temperature and sunlight also may impact pavement decay trends. Roads are also impacted by changing the number of freezing cycles as a consequence of climate change. These severe environmental conditions may accelerate pavement deterioration. Thus, in such a situation, an updated decay model capturing these changes can improve the quality of decision-making by reflecting reality.

## 6.2.5 *Travel Time*

Besides the safety and comfort of users and vehicle costs, improving pavement conditions would decrease travel time for road users. However, in classical PMS, travel time is not a typical and direct objective or maintenance intervention. Amador and Serrano (2017) proposed a methodology to consider the travel time to health centers, schools, workplaces, etc. in classical PMS.

The objective of this study was to incorporate travel time as a decision criterion to allocate a budget for maintenance and replacement treatments of pavement. The methodology was including the three-step process of:

- Estimate travel demand to hospitals and universities
- Characterize roads: Surface, alignment, condition
- Allocate interventions and upgrades to minimize travel time

The model defined a combined objective to minimize travel time considering demand and to maximize overall road performance (i.e., minimizing road roughness) as formulated below.

$$\text{Minimizing } Z = \sum\sum\sum \text{Total Travel Time} + \sum\sum\sum \text{Overall Roughness}$$

$$\text{Subject to}: \sum\sum\sum \text{Total Cost} \leq \text{Available Budget}$$

The model was applied to a real case study to objectively district budget and facilitate access to health and education centers improving life quality for rural areas (Fig.6.13).

### 6.2.6 *Vulnerability and Resiliency*

Measuring and incorporating the vulnerability of an asset to natural events is important to guide decisions on retrofits to increase the level of resiliency. The expected degree of damage (D) reflecting the structure vulnerability level can be used to expand traditional decision support systems.

Amin et al. (2020) proposed a new intervention-oriented platform to increase pavement resiliency as is shown in Table 6.3.

Only roads in poor condition at high and medium flooding risk zones are candidates for reconstruction as perpetual pavements. Lifecycle optimization to achieve and sustain good pavement conditions (i.e., decreasing network IRI) at a minimum cost is used to find required levels of the annual MRR budget.

$$\text{Minimize } Z = \sum\sum\sum \text{Total Cost}$$

$$\text{Subject to}: \sum\sum\sum \text{Overall } (\text{IRI}_t) \leq 0.9 \times \sum\sum\sum \text{Overall } (\text{IRI}_{t-1})$$

The minimization of roughness progression (IRI) and damage (D) values under such a budget is then used to find optimal strategic results for pavement management. This formulation relied on a transfer function that connects recursively all periods of time.

$$\text{Minimizing } Z = \sum\sum\sum \text{Overall Roughness} + \sum\sum\sum \text{Overall Damage}$$

$$\text{Subject to}: \sum\sum\sum \text{Total Cost} \leq \text{Available Budget (Stage I)}$$

The model was applied to a real case study and results are presented in Figs. 6.14 and 6.15. As can be seen, the proposed methodology was able to enhance both objectives of minimizing overall IRI and damage across the network.

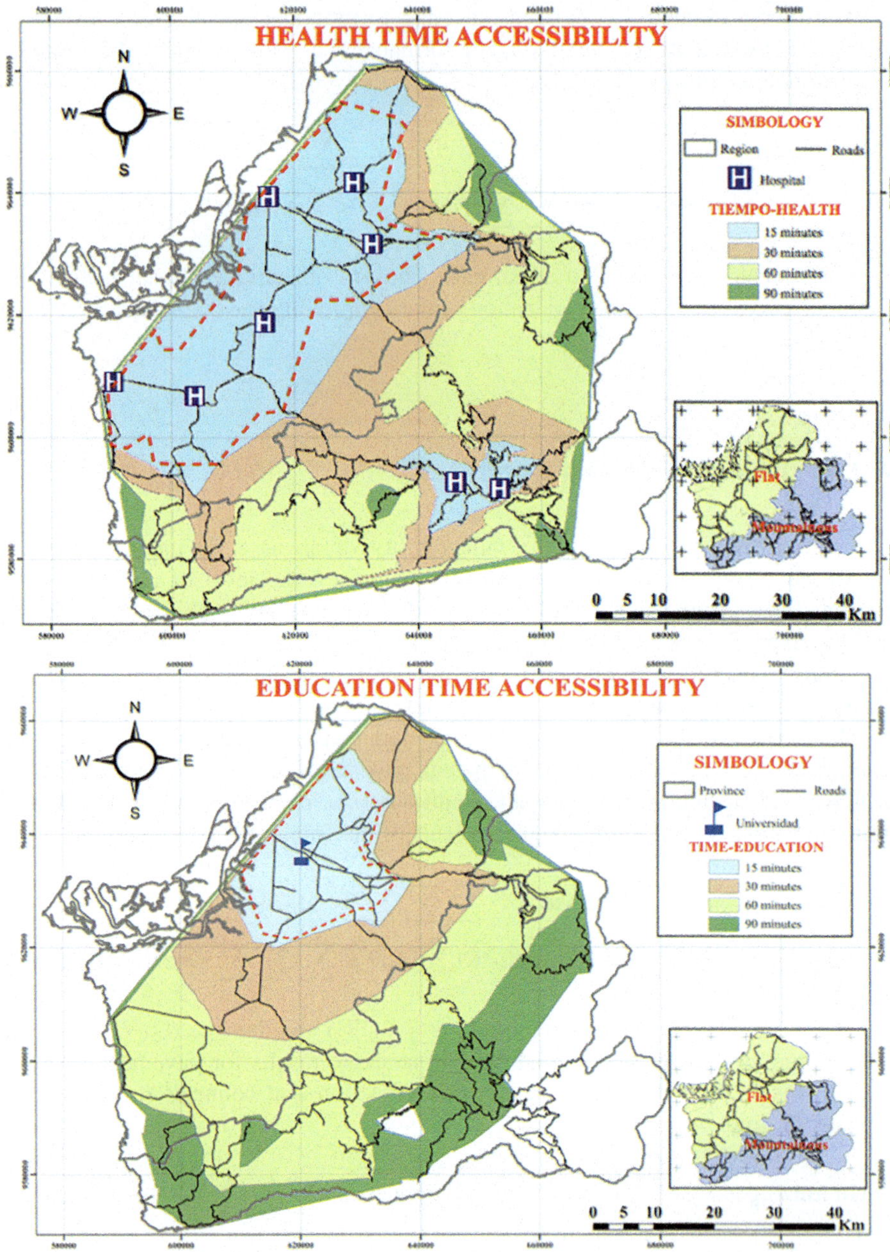


**Fig. 6.13** Education and Health time accessibility. (Amador and Serrano (2017))

**Table 6.3** Intervention to Increase the Resiliency of a Pavement

| Resilience Intervention | Road Hierarchy | GEOPHRIV[a] Zones | Operational Window | Cost (USD) per km |
|---|---|---|---|---|
| Reconstruction as a perpetual pavement with earth filling of about one meter plus the pavement structure. Protected with lateral walls. | Regional Highway | High | IRI > = 6 | 82234 |
| | | Medium | | |
| | Arterial | High | | |
| | | Medium | | |
| | Collector | High | | |
| | | Medium | | |

[a] GEO-PHysical RIsk and Vulnerability (GEOPHRIV)

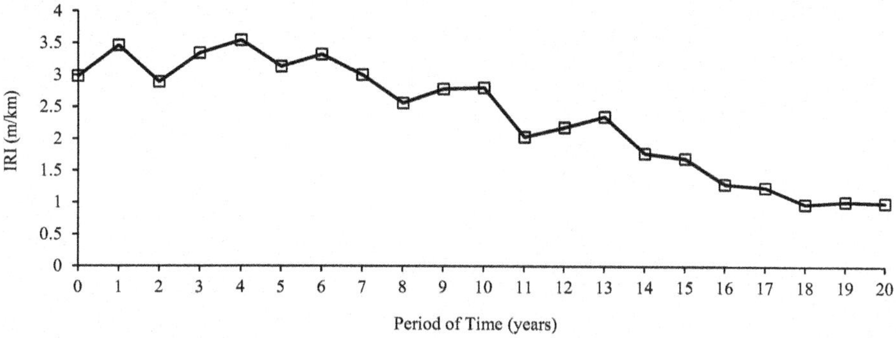


**Fig. 6.14** Network overall IRI. (Amin et al. (2018))

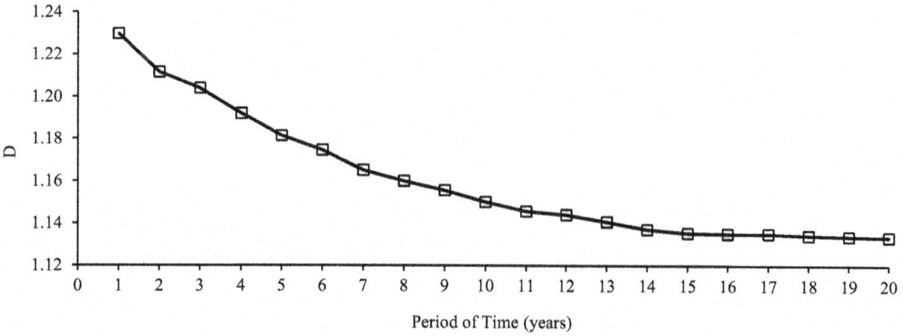


**Fig. 6.15** Network overall damage indicator (D). (Amin et al. (2018))

# References

Amador, L., Mohammadi, A., Abu-Samra, S., & Maghsoudi, R. (2020). Resilient storm pipes: A multi-stage decision support system. *Structure and Infrastructure Engineering, 16*(5), 847–859.

Amador, L., Mohammadi, A., & Nasiri, F. (2017). A cooperative environment to incorporate comfort on modal choices and trip assignment. In *ITS world congress, Oct 29-Nov 2*. Quebec.

Amador-Jimenez, L., & Serrano, L. (2017). Pavement management: A service-based optimal allocation of roads' interventions. *International Journal of Sustainable Development and Planning, 12*(6), 1096–1106.

Amin, M. S. R., Tamima, U., & Amador, L. (2020). Towards resilient roads to storm-surge flooding: Case study of Bangladesh. *International Journal of Pavement Engineering, 21*(1), 63–73.

Faghih-Imani, A., & Amador-Jimenez, L. (2013). Toward sustainable pavement management: Incorporating environmental impacts of pavement treatments into a performance-based optimization. *Transportation Research Record, 2366*(1), 13–21.

Mohammadi, A., Amador, L., & Nasiri, F. (2019). A multi-criteria assessment of the passengers' level of comfort in urban railway rolling stocks. *Journal of Sustainable Cities and Society, 53*(3), 101892.

Mohammadi, A., Elsaid, F., Amador-Jimenez, L., & Nasiri, F. (2018). Optimizing transit maintenance and rehabilitation to support human development and sustainability: A case study of Costa Rica's railroad network. *International Journal of Sustainable Transportation, 13*(7), 497–510. https://doi.org/10.1080/15568318.2018.1486488.

UN (United Nation) (2016) (b). Mobilizing sustainable transport for development. Retrieved from https://sustainabledevelopment.un.org/. [ 17 August 2017].

# Appendices

## Appendix 1

**Example A.1 Mechanistic-Empirical Pavement Rut Depth**
This example comes from the Mechanistic-Empirical Pavement Rut Depth (MEPDG) equation (which is estimated for each layer separately) and the traditional asphalt institute maximum number of repetitions of a given load to reach fatigue in terms of the rutting (0.5 inches or 12.7 mm). Structural rutting happens on each layer of the pavement and is the accumulated effect of all of them. To make it simple, consider two-layer pavement with 4 inches of HMA and 8 inches of well-compacted gravel. Rutting will then be measured at the mid-depth of each layer (depth variable) under temperature T and accumulated ESALs (N).

$$\text{PD} = \sum_{i=1}^{n} \varepsilon_{\text{p}}^{i} h^{i} \text{ where for asphalt layer the } \varepsilon_{\text{p}} = \left(k_1 10^{-3.4488} T^{1.5606} N^{0.479244}\right) \varepsilon_{\text{v}} \tag{A.1}$$

$$\varepsilon_{\text{v}} = \text{Vertical Elastic Strain in AC layer} \tag{A.2}$$

$$\varepsilon_{\text{v}} = \frac{1}{E}\left[\sigma_{\text{z}} - \upsilon(\sigma_{\text{r}} + \sigma_{\text{t}})\right] \tag{A.3}$$

Being $\upsilon$ the Poisson ratio (assume 0.35), $\sigma_{\text{z}}$ the vertical stress, $\sigma_{\text{t}}$ tangential stress and $\sigma_{\text{r}}$ radial stress, from which $\sigma_{\text{z}}$ has the largest value at the axis of symmetry of the load application, use of computer software is common to estimate the various stresses and build up the vertical strain; however, for the purpose of this book we will use a simplified equation for the estimation of the vertical elastic strain in the AC layer.



A. Mohammadi, L. Amador Jimenez, *Asset Management Decision-Making For Infrastructure Systems*, https://doi.org/10.1007/978-3-030-97614-9

$$\varepsilon_{\mathrm{v}} = \frac{(1+\nu)q}{E}\left[1 - 2\nu + \frac{2\nu z}{(a^2+z^2)^{0.5}} - \frac{z^3}{(a^2+z^2)^{1.5}}\right] \tag{A.4}$$

From which $q$ is the uniform distributed load from a 9000 lb. of an 18,000 lb. standard axle, $E$ is the elastic modulus of the asphalt cement material, which for this example is assumed as 300,000 psi (this value should come from a laboratory test), $z$ is the depth of the HMA layer—in this case, 6 inches—a is the radius of application of the dual tire ($a$), which under normal inflation (0.72 Mpa) circumstances is 6 inches for a 113.1 in$^2$ circular contact area produced with at least 4260 lbs. per tire on a dual tire with center to center for dual axles of 1.25–1.31 m (51 inches). For a Hac = 6, Base = 10 inch, and Subbase = 12 inch, and with those values, the equation reduces to:

$$\varepsilon_{\mathrm{v\ for\ ac-layer}} = \frac{(1+0.35)*79.57}{300,000}\left[1 - 2*0.35 + \frac{2*0.35*6}{(6^2+6^2)^{0.5}} - \frac{6^3}{(6^2+6^2)^{1.5}}\right]$$
$$= 0.000158$$

$$\varepsilon_{\mathrm{v\ for\ base-layer}} = \frac{(1+0.50)*79.57}{300,000}$$
$$\times\left[1 - 2*0.50 + \frac{2*0.35*6}{(6^2+16^2)^{0.5}} - \frac{6^3}{(6^2+16^2)^{1.5}}\right]$$
$$= 0.000461$$

$$\varepsilon_{\mathrm{v\ for\ subbase-layer}} = \frac{(1+0.50)*79.57}{300,000}$$
$$\times\left[1 - 2*0.50 + \frac{2*0.35*6}{(6^2+28^2)^{0.5}} - \frac{6^3}{(6^2+28^2)^{1.5}}\right]$$
$$= 0.000243$$

With all those values, we can replace back $\varepsilon_{\mathrm{v}}$ on the $\varepsilon_{\mathrm{p}}$ equation and explicitly consider the effect of temperature, accumulate ESALS (N) following the prediction of traffic loading. Having $\varepsilon_{\mathrm{p}}$ multiplying by the thickness of the layer, plastic deformation of the layer can be estimated as follows:

$$\text{Plastic deformation} = \varepsilon_{\mathrm{p}} h \ \text{ for each layer}$$

One detail is left pending, the value of $k_1$ which is estimated as follows:

$$k_1 = (C_1 + C_2 * \text{depth}) * 0.328196^{\text{depth}} \tag{A.5}$$

Where:

$$C_1 = -0.1039h_{ac}^2 + 2.4868h_{ac} - 17.342 \quad (A.6)$$

and

$$C_2 = -0.0172h_{ac}^2 - 1.7331h_{ac} + 27.42 \quad (A.7)$$

The depth to be used will be the thickness of the asphalt cement layer. Is it worth having a load restriction during the hottest months of the year in the middle east countries to avoid rutting deterioration of roads?

The rut depth of the granular layers is given in terms of the Water content (Wc) after 1 and $10^9$ load cycles at the resilient strain level $\varepsilon_r$ (recoverable strain when a load is retired) using the following set of equations for the plastic strain (ep) in terms of the vertical elastic strain $\varepsilon_v$.

$$\varepsilon_p = \left(\beta_G \frac{\varepsilon_0}{\varepsilon_r} e^{-\left(\frac{\rho}{N}\right)\beta}\right)\varepsilon_v \quad (A.8)$$

$$\beta_G = 1.673 \text{ or } 1.35 \text{ for base and subbase layers} \quad (A.9)$$

$$\frac{\varepsilon_0}{\varepsilon_r} = \text{weighted average of laboratory measurements after 1 and } 10^9 \text{ load cycles} \quad (A.10)$$

$$\frac{\varepsilon_0}{\varepsilon_r} = \frac{1}{2}\left[0.15\, e^{\left(\frac{\rho}{1}\right)^\beta} + 20e^{\left(\frac{\rho}{10^9}\right)^\beta}\right] \quad (A.11)$$

$$\beta = e^{-0.6119-0.017638\ Wc} \text{ and } \rho = 10^9\left[\frac{-4.89285}{1-\left(10^9\right)^\beta}\right]^\beta \quad (A.12)$$

For the given conditions, Table A.1 provides a summary of the rut depth design criterion for suggested accumulated ESALs (N) taken from field observations.

**Example A.2 Mechanistic-Empirical Pavement Cracking**

This example of the mechanistic-empirical equation is for cracking. The number of repetitions to failure $N_f$ an asphalt layer of thickness $h_{ac}$ with elastic modulus $E$ and experiencing tensile strains $\varepsilon_t$ under given loads and axle imprint is:

$$N_f = 0.00432 * k_1 C\left(\frac{1}{\varepsilon_t}\right)^{3.9492}\left(\frac{1}{E}\right)^{1.281} \quad (A.13)$$

$$K_{1\ \text{bottom up}} = \frac{1}{0.000398 + \frac{0.003602}{1+e^{11.02-3.49\ h_{ac}}}} \quad (A.14)$$

**Table A.1** Rut depth (permanent deformation) on AC, base, and subbase layers

| AC | | 3% growth | | | Base | 10 | BG = | **1.673** | | inches | Sub | 12 | BG = | **1.35** | | inches |
|---|---|---|---|---|---|---|---|---|---|---|---|---|---|---|---|---|
| hac | T | N | k1 | **rut_ac (inch)** | WC % | beta | rho | e0_er | e_P | **$rut_{base}$ = e_P * $h_{base}$** | WC % | beta | rho | e0_er | e_P | **$rut_{subbase}$ = e_P * $h_{subbase}$** |
| 9 | 75 | 3,000,000 | 0.37 | **0.2026** | 15 | 0.42 | 45.37 | 11.51 | 0.009 | **0.0879** | 20 | 0.381 | 64.53 | 11.52 | 0.003723 | **5.03E-03** |
| 9 | 75 | 3,182,700 | 0.37 | **0.2084** | 15 | 0.42 | 45.37 | 11.51 | 0.009 | **0.0879** | 20 | 0.381 | 64.53 | 11.52 | 0.003725 | **5.03E-03** |
| 9 | 75 | 3,278,181 | 0.37 | **0.2114** | 15 | 0.42 | 45.37 | 11.51 | 0.009 | **0.0879** | 20 | 0.381 | 64.53 | 11.52 | 0.003725 | **5.03E-03** |
| 9 | 75 | 3,376,526 | 0.37 | **0.2144** | 15 | 0.42 | 45.37 | 11.51 | 0.009 | **0.0879** | 20 | 0.381 | 64.53 | 11.52 | 0.003726 | **5.03E-03** |
| 9 | 75 | 3,477,822 | 0.37 | **0.2174** | 15 | 0.42 | 45.37 | 11.51 | 0.009 | **0.0879** | 20 | 0.381 | 64.53 | 11.52 | 0.003727 | **5.03E-03** |
| 9 | 75 | 3,582,157 | 0.37 | **0.2205** | 15 | 0.42 | 45.37 | 11.51 | 0.009 | **0.0879** | 20 | 0.381 | 64.53 | 11.52 | 0.003727 | **5.03E-03** |
| 9 | 75 | 3,689,622 | 0.37 | **0.2237** | 15 | 0.42 | 45.37 | 11.51 | 0.009 | **0.0879** | 20 | 0.381 | 64.53 | 11.52 | 0.003728 | **5.03E-03** |
| 9 | 75 | 3,800,310 | 0.37 | **0.2269** | 15 | 0.42 | 45.37 | 11.51 | 0.009 | **0.0879** | 20 | 0.381 | 64.53 | 11.52 | 0.003729 | **5.03E-03** |
| 9 | 75 | 3,914,320 | 0.37 | **0.2301** | 15 | 0.42 | 45.37 | 11.51 | 0.009 | **0.088** | 20 | 0.381 | 64.53 | 11.52 | 0.003729 | **5.03E-03** |
| 9 | 75 | 4,031,749 | 0.37 | **0.2334** | 15 | 0.42 | 45.37 | 11.51 | 0.009 | **0.088** | 20 | 0.381 | 64.53 | 11.52 | 0.003730 | **5.04E-03** |
| 9 | 75 | 4,152,702 | 0.37 | **0.2367** | 15 | 0.42 | 45.37 | 11.51 | 0.009 | **0.088** | 20 | 0.381 | 64.53 | 11.52 | 0.003730 | **5.04E-03** |
| 9 | 75 | 4,277,283 | 0.37 | **0.2401** | 15 | 0.42 | 45.37 | 11.51 | 0.009 | **0.088** | 20 | 0.381 | 64.53 | 11.52 | 0.003731 | **5.04E-03** |
| 9 | 75 | 4,405,601 | 0.37 | **0.2435** | 15 | 0.42 | 45.37 | 11.51 | 0.009 | **0.088** | 20 | 0.381 | 64.53 | 11.52 | 0.003732 | **5.04E-03** |
| 9 | 75 | 4,537,769 | 0.37 | **0.247** | 15 | 0.42 | 45.37 | 11.51 | 0.009 | **0.088** | 20 | 0.381 | 64.53 | 11.52 | 0.003732 | **5.04E-03** |
| 9 | 75 | 4,673,902 | 0.37 | **0.2505** | 15 | 0.42 | 45.37 | 11.51 | 0.009 | **0.088** | 20 | 0.381 | 64.53 | 11.52 | 0.003733 | **5.04E-03** |
| 9 | 75 | 4,814,119 | 0.37 | **0.2541** | 15 | 0.42 | 45.37 | 11.51 | 0.009 | **0.088** | 20 | 0.381 | 64.53 | 11.52 | 0.003733 | **5.04E-03** |
| 9 | 75 | 4,958,543 | 0.37 | **0.2577** | 15 | 0.42 | 45.37 | 11.51 | 0.009 | **0.088** | 20 | 0.381 | 64.53 | 11.52 | 0.003734 | **5.04E-03** |
| 9 | 75 | 5,107,299 | 0.37 | **0.2614** | 15 | 0.42 | 45.37 | 11.51 | 0.009 | **0.088** | 20 | 0.381 | 64.53 | 11.52 | 0.003735 | **5.04E-03** |
| 9 | 75 | 5,260,518 | 0.37 | **0.2651** | 15 | 0.42 | 45.37 | 11.51 | 0.009 | **0.088** | 20 | 0.381 | 64.53 | 11.52 | 0.003735 | **5.04E-03** |
| 9 | 75 | 5,418,334 | 0.37 | **0.2689** | 15 | 0.42 | 45.37 | 11.51 | 0.009 | **0.0881** | 20 | 0.381 | 64.53 | 11.52 | 0.003736 | **5.04E-03** |

**Table A.2** Estimation of contact pressure for various axle loads and axle types

| Radius | Single KN | Single (lbs) | Contact pressure (psi) | Tandem KN (lbs) | Tandem (lbs) | Contact pressure (psi) | Dual space (mm) | Dual space (inch) |
|---|---|---|---|---|---|---|---|---|
| 6 | 133 | 29898.4 | 264.36 | 231 | 51928.8 | 459.15 | 318 | 12.520 |
| 6 | 125 | 28,100 | 248.46 | 213 | 47882.4 | 423.37 | 318 | 12.520 |
| 6 | 115 | 25,852 | 228.58 | 195 | 43,836 | 387.60 | 318 | 12.520 |
| 6 | 107 | 24053.6 | 212.68 | 178 | 40014.4 | 353.80 | 318 | 12.520 |

$$K_{1\text{ top down}} = \frac{1}{0.01 + \frac{12}{1+e^{15.676-2.8186\,h_{ac}}}} \tag{A.15}$$

At this point, the reader must notice that the modulus of elasticity of the asphalt changes due to temperature variations which can be taken on a monthly ($i$) and hourly ($j$) basis. The modulus of elasticity of the base and subbase also change at a monthly ($i$) basis due to moisture variations. Similarly, the tensile strain varies per axle type ($k$), i.e., single, tandem, triple, and quad axles, and depending on the load level on each axle ($l$) and the path of load application ($m$), which can be assumed normally distributed lateral wheel wander. For this reason, the estimation of the number of applied load applications ($n_{i,\ j,\ k,\ l,\ m}$) is done for each of the previous conditions and compared to the allowable (maximum) number of axle applications to cracking failure using the $N_f$ for the same conditions.

$$\text{FD} = \text{Fatigue Damage (percentage on decimal base)} = \sum \frac{n_{i,j,k,l,m}}{N_{i,j,k,l,m}} \tag{A.16}$$

and

$$\text{Area of Cracking (alligator)} = \frac{1560}{1 + e^{(7-3.5*\log \text{FD})}} \tag{A.17}$$

Let us look at a simple example to illustrate the method where we ignore hourly variations. The contract pressure of the loads experienced for single and dual axles is shown in Table A.2. Table 3.4 provides the monthly variation of average temperatures, load spectrum of single and tandem axle loads, and resulting tensile strains calculated with the KENPAVE software suite (Huang, 2003).

We will test the suitability of a 150 mm asphalt layer (6 inches) for bottom-up (alligator) cracking. The K1 value for bottom-up is = 250.01 for this example. We calculate the tensile strain $\varepsilon_t$ for each combination case of the previous table and estimate the corresponding allowable ($N_f$) as follows.

The analysis of the allowable number of loads is given by $N_f = 0.00432 * k_1 C\left(\frac{1}{\varepsilon_t}\right)^{3.9492}\left(\frac{1}{E}\right)^{1.281}$ with $K_1 = 250$ for each condition of Table A.3 before. For

**Table A.3** Moduli, loads and tensile strains

| Month ($i$) | $E$ MPa (psi) | E granular | Single KN (lbs) | $\varepsilon_t$ single | Tandem KN (lbs) | $\varepsilon_t$ tandem |
|---|---|---|---|---|---|---|
| Dec., Jan, Feb | 2000 (290,000) | 145 (21050) | 133 (29900) | 1.25E-03 | 231 (51931) | 2.45E-03 |
| Dec., Jan, Feb | 2000 (290,000) | 145 (21050) | 125 (28101) | 1.17E-03 | 213 (47884) | 2.23E-03 |
| Dec., Jan, Feb | 2000 (290,000) | 145 (21050) | 115 (25853) | 1.08E-03 | 195 (43837) | 2.03E-03 |
| Dec., Jan, Feb | 2000 (290,000) | 145 (21050) | 107 (24054) | 1.00E-03 | 178 (40016) | 1.85E-03 |
| March, April, May | 1800 (261,000) | 150 (21750) | 133 (29900) | 1.24E-03 | 231 (51931) | 2.45E-03 |
| March, April, May | 1800 (261,000) | 150 (21750) | 125 (28101) | 1.17E-03 | 213 (47884) | 2.22E-03 |
| March, April, May | 1800 (261,000) | 150 (21750) | 115 (25853) | 1.08E-03 | 195 (43837) | 2.03E-03 |
| March, April, May | 1800 (261,000) | 150 (21750) | 107 (24054) | 1.01E-03 | 178 (40016) | 1.85E-03 |
| June, July, August | 1500 (217,500) | 175 (25375) | 133 (29900) | 1.25E-03 | 231 (51931) | 2.41E-03 |
| June, July, August | 1500 (217,500) | 175 (25375) | 125 (28101) | 1.18E-03 | 213 (47884) | 2.22E-03 |
| June, July, August | 1500 (217,500) | 175 (25375) | 115 (25853) | 1.08E-03 | 195 (43837) | 2.03E-03 |
| June, July, August | 1500 (217,500) | 175 (25375) | 107 (24054) | 1.01E-03 | 178 (40016) | 1.85E-03 |
| Sept., Oct., Nov. | 1750 (253,750) | 175 (25375) | 133 (29900) | 1.25E-03 | 231 (51931) | 2.41E+03 |
| Sept., Oct., Nov. | 1750 (253,750) | 175 (25375) | 125 (28101) | 1.18E-03 | 213 (47884) | 2.22E-03 |
| Sept., Oct., Nov. | 1750 (253,750) | 175 (25375) | 115 (25853) | 1.08E-03 | 195 (43837) | 2.03E-03 |
| Sept., Oct., Nov. | 1750 (253,750) | 175 (25375) | 107 (24054) | 1.07E-03 | 178 (40016) | 1.85E-03 |

the $C$ constant, we use 11% as asphalt binder volume and 5% as air void volume into the following equation: $C = 10^{4.84\left(\frac{V_b}{V_a+V_b}-0.69\right)}$ =0.97252

The final calculation of cracking for the observed n single and tandem load repetitions for each group of months and as per estimations made using travel demand modeling or trip generation rates are presented in Table A.4. As the reader can notice, this model is quite complex and requires mechanistic estimations of tensile strains using elastic theory analysis. For this reason, the use of mechanistic equations is sometimes ignored in the development of performance models for the condition.

**Table A.4** Estimation of observed (expected) load repetitions for each group and cracking life

| Month (i) | E MPa (psi) | Tensile strain single | Tensile strain tandem | Allowable Nf single | Allowable Nf tandem | Observed n single | Observed n tandem | n/Nf single | n/Nf tandem |
|---|---|---|---|---|---|---|---|---|---|
| December, January, February | 290,000 | 1.25E-03 | 2.45E-03 | 30,922 | 2161 | 300 | 200 | 0.009702 | 0.092537 |
| December, January, February | 290,000 | 1.17E-03 | 2.23E-03 | 40,025 | 3134 | 250 | 150 | 0.006246 | 0.047864 |
| December, January, February | 290,000 | 1.08E-03 | 2.03E-03 | 54,905 | 4542 | 568 | 125 | 0.010345 | 0.027521 |
| December, January, February | 290,000 | 1.00E-03 | 1.85E-03 | 74,406 | 6554 | 800 | 450 | 0.010752 | 0.068662 |
| March, April, and May | 261,000 | 1.24E-03 | 2.45E-03 | 36,415 | 2474 | 400 | 350 | 0.010984 | 0.141494 |
| March, April, and May | 261,000 | 1.17E-03 | 2.22E-03 | 45,808 | 3677 | 1000 | 250 | 0.02183 | 0.067989 |
| March, April, and May | 261,000 | 1.08E-03 | 2.03E-03 | 62,609 | 5219 | 1500 | 50 | 0.023958 | 0.009581 |
| March, April, and May | 261,000 | 1.01E-03 | 1.85E-03 | 83,170 | 7501 | 1800 | 75 | 0.021643 | 0.009999 |
| June, July, and August | 217,500 | 1.25E-03 | 2.41E-03 | 44,700 | 3362 | 2000 | 85 | 0.044742 | 0.025285 |
| June, July, and August | 217,500 | 1.18E-03 | 2.22E-03 | 56,893 | 4644 | 2100 | 100 | 0.036911 | 0.021531 |
| June, July, and August | 217,500 | 1.08E-03 | 2.03E-03 | 79,081 | 6591 | 2500 | 110 | 0.031613 | 0.016688 |
| June, July, and August | 217,500 | 1.01E-03 | 1.85E-03 | 105,050 | 9474 | 3000 | 150 | 0.028558 | 0.015833 |
| September, October, November | 253,750 | 1.25E-03 | 2.41E-03 | 36,690 | 2759 | 1000 | 90 | 0.027255 | 0.032617 |
|  | 253,750 | 1.18E-03 | 2.22E-03 | 46,699 | 3812 | 500 | 80 | 0.010707 | 0.020985 |

(continued)

**Table A.4** (continued)

| Month (i) | E MPa (psi) | Tensile strain single | Tensile strain tandem | Allowable Nf single | Allowable Nf tandem | Observed n single | Observed n tandem | n/Nf single | n/Nf tandem |
|---|---|---|---|---|---|---|---|---|---|
| September, October, November | | | | | | | | | |
| September, October, November | 253,750 | 1.08E-03 | 2.03E-03 | 64,910 | 5410 | 250 | 40 | 0.003851 | 0.007393 |
| September, October, November | 253,750 | 1.07E-03 | 1.85E-03 | 68,593 | 7776 | 150 | 20 | 0.002187 | 0.002572 |
| | | | | | | | Subtotals | 0.301285 | 0.608552 |
| | | | | | Aligator fatigue | Cracking | Total n/Nf | 0.909837 | 90.98% |

The previous example shows that the example pavement will arrive at a 90.98% FC, that is, the pavement provided 6 inches and material characteristics are expected to perform well during the service life accumulating total damage or nearly 91% at the end of its lifespan.

## *Stochastic Estimation with Empirical Deterioration*

An empirical equation linking the desired deterioration response with the available contributing (independent) factors (variables) is used to estimate stochastically the calibration coefficients for a given homogeneous group. For example, consider the simplified IRI equation proposed by Patterson and Attoh-Okine (1992) (Eq. 3.1), which predicts the progression of IRI where alpha is the calibration parameter for the demand/capacity ratio (alpha*[NESALS / (1 + SNC)]^5) using the open-source software OpenBUGS.

Imagine you count with the following characteristics for a pavement taken from 10 years of observations of NESALs and IRI and that the province extends over 2 climatic regions (Table A.5).

A Full Bayesian Markov Chain Monte Carlo simulation approach can be used to calibrate the coefficients (alpha and beta) with the dataset (Amador Jimenez & Mrawira, 2011). In OpenBUGS, you must define the functional form, the stochastic nodes (alpha and beta), and use the dataset to learn (estimate) the coefficients. Notice how having two climatic regions and presuming their effect is not significant leads us to propose one model for two homogeneous groups together indexing the alpha and beta in the model with the z-letter for environmental regions and the i-letter for each observation on the dataset. Stochastic nodes are indexed with the r-letter for the environmental region.

**Table A.5** Sample data for model

| M | IRI | Age | NEM (in millions) | Area | Climate region (*z*)[a] |
|---|---|---|---|---|---|
| 0.07 | 1.91 | 2 | 7.857234 | 23.6 | 1 |
| 0.07 | 1.96 | 1 | 7.857234 | 24.05 | 1 |
| 0.07 | 1.97 | 2 | 7.857234 | 25.69 | 1 |
| 0.07 | 2 | 2 | 7.857234 | 24.55 | 1 |
| ... | ... | ... | ... | ... | ... |
| 0.074 | 3.3 | 13 | 6.011766 | 22.01 | 2 |
| 0.074 | 2.45 | 11 | 6.115438 | 27.19 | 2 |
| 0.074 | 1.13 | 8 | 6.906708 | 26.26 | 2 |
| 0.074 | 1.36 | 10 | 8.627733 | 25.25 | 2 |
| 0.074 | 1.32 | 10 | 8.831204 | 24.53 | 2 |
| ... | ... | ... | ... | ... | ... |

[a]Note: For OpenBUGS, the dataset must be indexed to reflect the groups (in this case climatic regions)

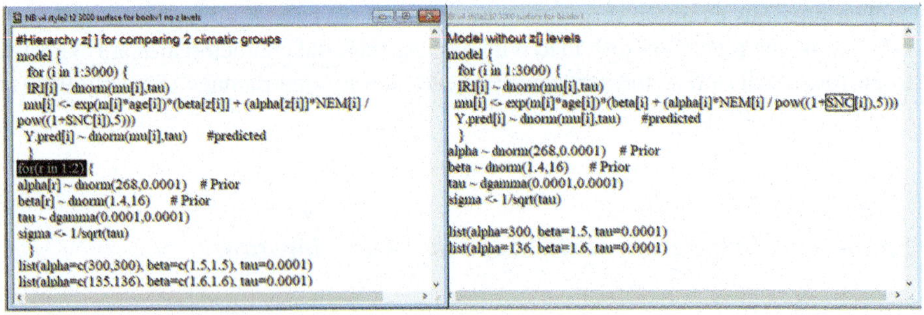


**Fig. A.1** Stochastic model with and without hierarchical "levels" to compare climatic groups

A Full Bayesian Markov Chain Monte Carlo simulation approach is to be used to calibrate the coefficients (alpha and beta) with the dataset. In OpenBUGS, you must define the functional form, the stochastic nodes (alpha and beta), and use the dataset to learn (estimate) the coefficients. Notice how having two climatic regions and presuming their effect is not significant leads us to propose one model for two homogeneous groups together indexing the alpha and beta in the model with the z-letter for environmental regions and the i-letter for each observation on the dataset. Stochastic nodes are indexed with the r-letter for the environmental region (Fig. A.1).

Use of proxy values: A model calibrated stochastically

Imagine now you have all values necessary for the previous mechanistic-empirical model but the SNC is not available at all. A proxy measure called deflection area basin (in short "area") can be used instead of SNC. The proxy replaces SNC in the mechanistic-empirical equation where we predict the progression of IRI. In this case, we estimate alpha and beta as parameters—the initial value of IRI and the contribution to deterioration from the demand/capacity ratio as given by the ratio of accumulated ESALS / area^5.

$$\mathrm{IRI} = e^{m*t} * \left[\beta + \alpha * \left(\frac{\mathrm{NESALS}}{(\mathrm{area})^5}\right)\right] \tag{A.18}$$

The OpenBUGS model for this example is created to use a dataset of 3000 observations from which all values of deflection area basin ("area"), accumulated ESALs (NEM), and environmental moisture index coefficient ("m") are available. Once again, the model could take advantage of hierarchical levels to make comparisons across homogeneous groups. The model in Fig. A.2 can be used to compare two climatic regions.

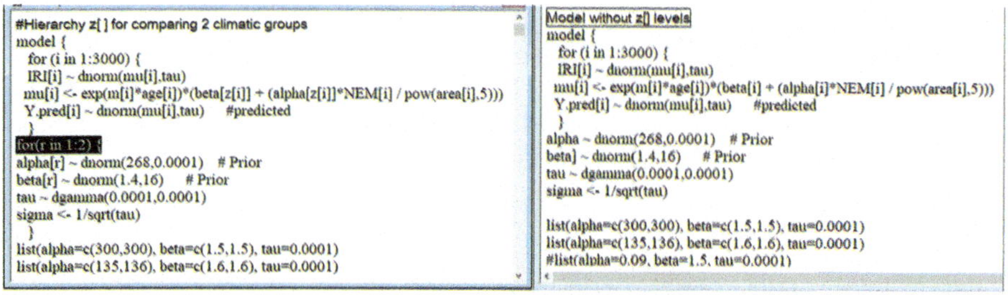


**Fig. A.2** Stochastic model with "area" as a proxy for SNC, with and without hierarchical levels z[]

## *Imputation Through Stochastic*

The use of stochastics nodes is advantageous when the dataset is missing some values. Stochastics nodes are syntactic elements defined using the mean and the standard deviation of the actual values observed for a given independent variable (IV). For any IV missing value, the stochastic node randomly generates an RV in a way that does not bias the probabilistic distribution of the IV. The use of stochastic nodes could theoretically be employed for a given asset/segment where all IV values are available but the dependent variable is missing. The use of stochastic nodes is encouraged when a large portion of the dataset will otherwise be removed. The inclusion of such a portion of the dataset aids to accomplish the law of large numbers.

Let us now consider an example: Imagine we have 482 segments missing values of the deflection basin area (a characteristic depicting the structural strength) for a given pavement. Once again we recur to OpenBUGS freeware, in which a stochastic node for the missing Deflection Area Basin ("area") values is created assuming the distribution of SN follows a normal distribution and estimating the mean is 25 and the standard deviation(sd) is 5.5049 (Precision = 1/sd^2 = 0.033). The stochastic node is called SN and is indexed by the letter p; therefore OpenBUGS is defined as a loop for the 482 missing points.

Every time a missing value of SN appears, the software will recur to the *area(p)* node and draw a value that is then used to replace the missing value. The updated OpenBUGS model is shown in Fig. A.3.

## *Stochastic Estimation with Mechanistic-Empirical*

To implement the rutting MEPDG model (Example A.1), the model is used to calibrate to local condition alpha and beta coefficients from the AC_rut depth equation. Let us implement in OpenBUGS the previous equation for the AC layer only (Fig. A.4).

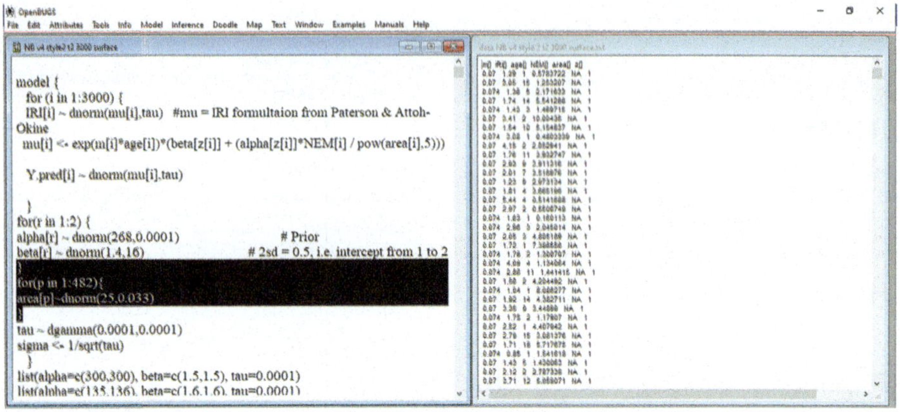


**Fig. A.3** OpenBUGS for imputation through stochastic model

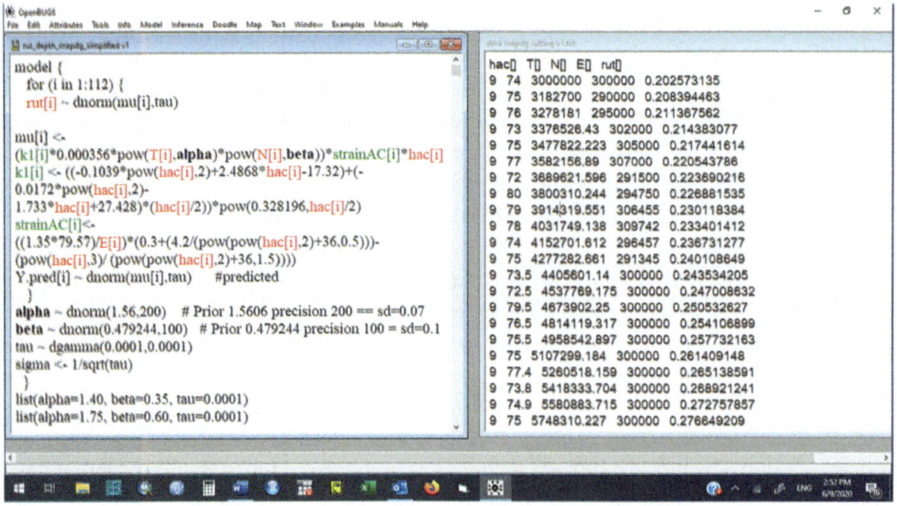


**Fig. A.4** Example of MEPDG rut depth (permanent deformation) model for the AC layer only

The estimated alpha and beta with OpenBUGS for the local observations are shown in Fig. A.5 below.

To implement the cracking MEPDG formulation in OpenBUGS, we need to rely on the conversion of separate axle loads back into ESALs; the reader should notice that the design of pavements should take full advantage of the axle load spectra enabled by the NCHRP (2004). For this example; however, we need to rely on the simplification to ESALs to facilitate the illustration and avoid the summation across axle types.

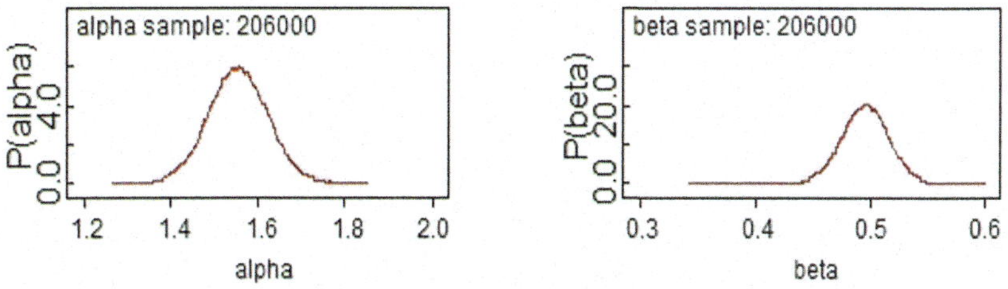


| | mean | sd | MC_error | val2.5pc | median | val97.5pc | start | sample |
|---|---|---|---|---|---|---|---|---|
| alpha | 1.552 | 0.0673 | 6.179E-4 | 1.422 | 1.552 | 1.685 | 1 | 206000 |
| beta | 0.4958 | 0.01988 | 1.845E-4 | 0.4565 | 0.4959 | 0.5341 | 1 | 206000 |

**Fig. A.5** MEPDG rutting model results—AC layer

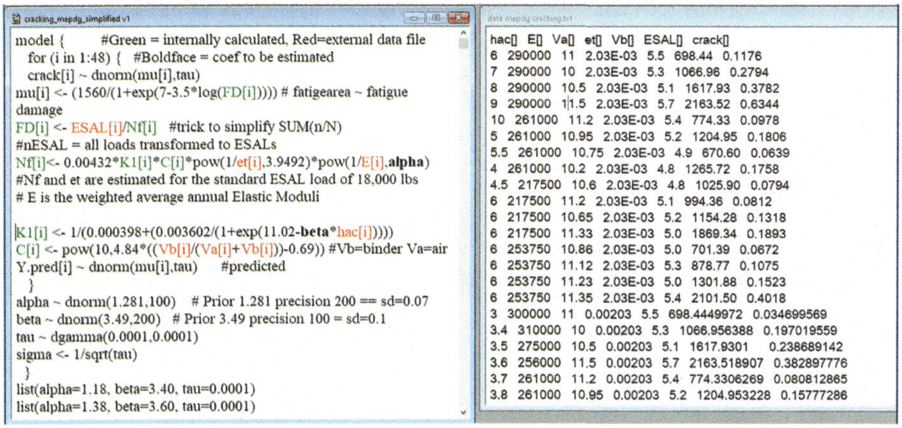

cracking_mepdg_simplified v1

```
model {        #Green = internally calculated, Red=external data file
  for (i in 1:48) {   #Boldface = coef to be estimated
  crack[i] ~ dnorm(mu[i],tau)
mu[i] <- (1560/(1+exp(7-3.5*log(FD[i])))) # fatigearea ~ fatigue
damage
FD[i] <- ESAL[i]/Nf[i]   #trick to simplify SUM(n/N)
#nESAL = all loads transformed to ESALs
Nf[i]<- 0.00432*K1[i]*C[i]*pow(1/et[i],3.9492)*pow(1/E[i],alpha)
#Nf and et are estimated for the standard ESAL load of 18,000 lbs
# E is the weighted average annual Elastic Moduli

K1[i] <- 1/(0.000398+(0.003602/(1+exp(11.02-beta*hac[i]))))
C[i] <- pow(10,4.84*((Vb[i]/(Va[i]+Vb[i]))-0.69)) #Vb=binder Va=air
Y.pred[i] ~ dnorm(mu[i],tau)     #predicted
  }
alpha ~ dnorm(1.281,100)    # Prior 1.281 precision 200 == sd=0.07
beta ~ dnorm(3.49,200)   # Prior 3.49 precision 100 = sd=0.1
tau ~ dgamma(0.0001,0.0001)
sigma <- 1/sqrt(tau)
  }
list(alpha=1.18, beta=3.40, tau=0.0001)
list(alpha=1.38, beta=3.60, tau=0.0001)
```

data mepdg cracking.txt

| hac[] | E[] | Va[] | et[] | Vb[] | ESAL[] | crack[] |
|---|---|---|---|---|---|---|
| 6 | 290000 | 11 | 2.03E-03 | 5.5 | 698.44 | 0.1176 |
| 7 | 290000 | 10 | 2.03E-03 | 5.3 | 1066.96 | 0.2794 |
| 8 | 290000 | 10.5 | 2.03E-03 | 5.1 | 1617.93 | 0.3782 |
| 9 | 290000 | 11.5 | 2.03E-03 | 5.7 | 2163.52 | 0.6344 |
| 10 | 261000 | 11.2 | 2.03E-03 | 5.4 | 774.33 | 0.0978 |
| 5 | 261000 | 10.95 | 2.03E-03 | 5.2 | 1204.95 | 0.1806 |
| 5.5 | 261000 | 10.75 | 2.03E-03 | 4.9 | 670.60 | 0.0639 |
| 4 | 261000 | 10.2 | 2.03E-03 | 4.8 | 1265.72 | 0.1758 |
| 4.5 | 217500 | 10.6 | 2.03E-03 | 4.8 | 1025.90 | 0.0794 |
| 6 | 217500 | 11.2 | 2.03E-03 | 5.1 | 994.36 | 0.0812 |
| 6 | 217500 | 10.65 | 2.03E-03 | 5.2 | 1154.28 | 0.1318 |
| 6 | 217500 | 11.33 | 2.03E-03 | 5.0 | 1869.34 | 0.1893 |
| 6 | 253750 | 10.86 | 2.03E-03 | 5.0 | 701.39 | 0.0672 |
| 6 | 253750 | 11.12 | 2.03E-03 | 5.3 | 878.77 | 0.1075 |
| 6 | 253750 | 11.23 | 2.03E-03 | 5.0 | 1301.88 | 0.1523 |
| 6 | 253750 | 11.35 | 2.03E-03 | 5.4 | 2101.50 | 0.4018 |
| 3 | 300000 | 11 | 0.00203 | 5.5 | 698.4449972 | 0.034699569 |
| 3.4 | 310000 | 10 | 0.00203 | 5.3 | 1066.956388 | 0.197019559 |
| 3.5 | 275000 | 10.5 | 0.00203 | 5.1 | 1617.9301 | 0.238689142 |
| 3.6 | 256000 | 11.5 | 0.00203 | 5.7 | 2163.518907 | 0.382897776 |
| 3.7 | 261000 | 11.2 | 0.00203 | 5.4 | 774.3306269 | 0.080812865 |
| 3.8 | 261000 | 10.95 | 0.00203 | 5.2 | 1204.953228 | 0.15777286 |

**Fig. A.6** Mechanistic-empirical: FD from alligator (Bottom-Up) cracking

First, we use Transport Canada (1993) equations to turn the observed axles into the equivalent standard.

$$\text{Single} = \frac{L^{2.9093}}{413.565}, \quad \text{andem} = \frac{L^{2.540}}{660.06}, \text{Tridem} = \frac{L^{2.113}}{423.19} \tag{A.19}$$

Second, we use the critical elastic tensile strain for an 18,000 lbs. axle and the weighted annual average elastic modulus for the asphalt to obtain the allowable number of ESALs in one year.

Third, we utilize a database containing observations of annual cracked area (alligator) for road segments to calibrate the coefficients for the elastic modulus of the AC (E) and the thickness of the hot mix asphalt layer (hac) (Figs. A.6 and A.7).

The estimated alpha and beta with OpenBUGs are shown below.

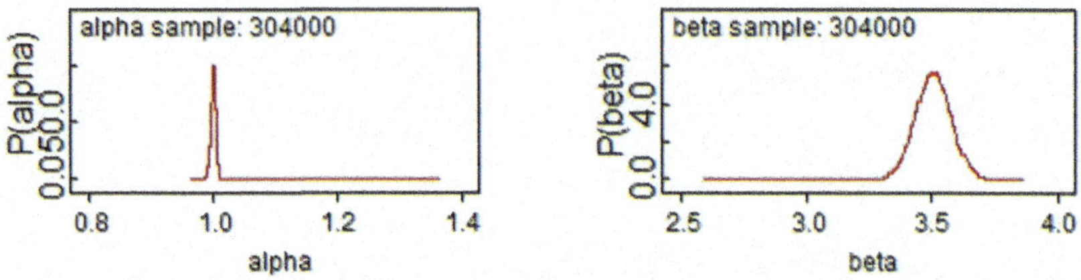


| | mean | sd | MC_error | val2.5pc | median | val97.5pc | start | sample |
|---|---|---|---|---|---|---|---|---|
| alpha | 1.002 | 0.004607 | 2.192E-5 | 0.9915 | 1.002 | 1.009 | 1 | 304000 |
| beta | 3.51 | 0.07394 | 7.45E-4 | 3.372 | 3.511 | 3.647 | 1 | 304000 |

**Fig. A.7** MEPDG alligator cracking model results

**Table A.6** Sample maximum service flow based on density ratios

| LOS | Density (pc/mi/ln) | Maximum V/C | Max service flow (pc/ph/pl) |
|---|---|---|---|
| A | <12 | 0.3 | 600 |
| B | <20 | 0.5 | 1000 |
| C | <28 | 0.7 | 1400 |
| D | <34 | 0.84 | 1670 |
| E | <43 | 1 | 2000 |
| F | >43 | – | – |

Taken from HCM (2000)

# Appendix 2

For instance, for roads, it is the observed number of passenger cars, per mile, per hour, per lane ($Q$) on a given road segment as compared to the maximum number of passenger cars that the road lane can handle per hour ($Q_{\max}$), and it had an impact on the operational speed ($V_{\text{cur}}$) by reducing the free-flow speed ($V_0$) as given by the following Eq. (A.20).

$$V_{\text{cur}} = \frac{V_0}{1 + a * \left(\frac{Q}{c * Q_{\max}}\right)^b} \tag{A.20}$$

Where $a$, $b$, and $c$ are calibration coefficients, which can be estimated from observed relationships for segments within the same type of road environment (i.e., multilane expressways, multi-lane arterials, two-lane collectors, etc.). The calibration requires the current observed speed and corresponding traffic volume.

The estimation of $Q_{\max}$ can be based on the density of vehicles and the corresponding maximum service flow as shown in Table A.6.

For densities beyond 43 passenger cars per mile per lane per hour, one can utilize the models of the *Highway Capacity Manual*; in this book and for illustration purposes, we simplify such models and use a simple ratio to illustrate the estimation of Qmax under saturation conditions.

$$Q_{\max} = 2000 * \left(1 - \frac{Q}{2000}\right) \tag{A.21}$$

The same happens for intersections where each of the approaches is delayed through a control device (traffic signal) controlled by the size of the queue of vehicles on each turning movement and the availability of turning lanes as opposed to lanes being shared by turning and straight movements.

For the transportation system in general, it means the number of trips generated (produced and attracted) by residences, commerce, industry, institutions, offices, entertainment sites, to mention a few. Entire land use classifications of over 100 types can be found in manuals such as the Institute of Transportation Engineering Trip Generation manual (ITE, 2017), the Transport Database (TDB, 2018) of Australia and New Zealand, the Transportation Information Computerized System of the United Kingdom (TRICS, 2018). Trip Generation can be used in isolation to gather an initial rough estimation of peak-time travel. More complex modeling approaches exist: they take advantage of four-step models to simulate travel demand, often from household interviews containing travel diaries, which are then turned into chains of origin-destination trips and aggregated across for a large enough sample that is representative of the area or country being studied. These origin-destination databases are then expanded using a synthetic population simulator that will clone households by type based on the future socio-economic forecast.

The product of a travel demand model is allocated on each mode of transportation (sidewalks, cycleways, roads, railways) and each link of a given network, using proximity criteria between the centroid of the geographic blocks where households are aggregated (called transportation analysis zones) and the closest link. These models produce the demand which is then compared to the supply (i.e. the capacity) of each link, which is easy to estimate based on the geometrical design and the design speed, and the frequency of service (Table A.7).

Systems such as TRANSCAD, EMME, VISUM, etc. are the best representatives of macro, meso, and microsimulation.

There is however the demand created by the movement of freight, which is different from the movement of people presented before. Freight depends on the degree of economic activities, industries producing goods and merchandises, which translate into trucks moving them in the network, trucks or trains delivering larger containers to ports (sea or air) where merchandises are exported, or collecting imported goods and doing the reverse: distributing back into the markets for final consumption or the intermediate industries for further transformation.

Input-output models are best suited to estimate trucking demand and changes in land uses from the transformation of floor space necessary to accommodate changes in the demand for new industries. They also connect well with employment because this ties back to the economic changes and the number and volume of goods produced by industries. Large tables of land use and industry types are necessary for each pair of spatial movements.

**Table A.7** Capacity and demand for public transportation options in Montreal (Sociéte de transport de Montréal 2012)

| System characteristic | Minibus | Bus | Articulated bus | Trolleybus | BRT | Tramway | Metro |
|---|---|---|---|---|---|---|---|
| Site (length) | 8–11 m | 12 m | 18 m | 18 m | 18 m | (1 unit = 1 car) 30–40 m | (1 train of 9 cars) 152 m |
| Seating capacity | 20 | 30 | 54 | 54 | 47 | 75 | 306 |
| Maximum capacity | 35 | 75 | 105 | 105 | 105 | 200 | 1200 |
| Passenger/hour/ peak Direction | 500–750 | 1000–1500 | 1500–2000 | 1500–2000 | 2000–3000 | 2000–4000 | 20,000–30,000 |
| Distance between stops | 250–500 m | 250–500 m | 250–500 m | 250–500 m | 400–500 m | 400–500 m between stops | 950 m between stops |
| Journey speed | 10–25 km/h | 10–25 km/h | 10–25 km/h | 10–25 km/h | 17–25 km/h | 17–25 km/h | 35–38 km/h |
| Service life | 12–16 years | 16 years | 16 years | 20–25 years | 16 years | 25 years | 40 years |
| Base unit cost per vehicle | $550,000–$ 670,000 (hybrid propulsion) | $470,000 (diesel propulsion) $900,000 (hybrid propulsion) | $700,000 (diesel propulsion) $900,000 (hybrid propulsion) | 1,000,000 | $2,000,000 (diesel propulsion) $3,000,000 (hybrid propulsion) | Approx. $2,000,000 | $2,000,000 to $3,000,000 |
| HG emissions (gCO2e/km) | approx,900 (hybrid propulsion) | 1453 (diesel propulsion) 1023 (hybrid propulsion) | 2099 (diesel propulsion) 1561 (hybrid propulsion) | None | 2099(diesel propulsion) 1561(hybrid propulsion) | None | None |

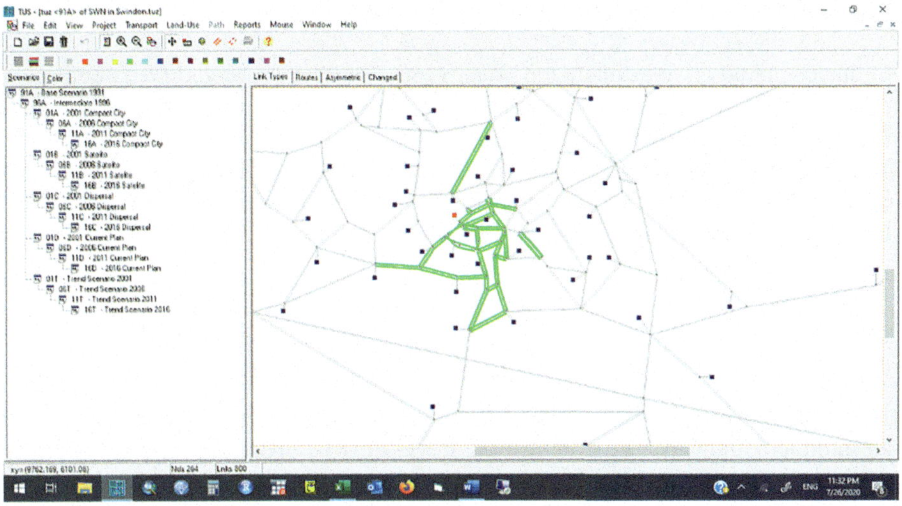

**Fig. A.8** Sample TRANUS Model for Swindon (de la Barra, 2014)

Systems such as TRANUS, PECAS, MEPLAN, and others are best suited to support the creation of travel demand that complements those of the households. At the time of writing this book, efforts were carried out to merge both types of systems (Fig. A.8).

Important to notice is that, as seen in this section, land use development is of utmost importance to forecast demand. Land use changes reflect economic and social development. Think for instance in nations that are developing fast; they count with large numbers of laborers coming from overseas; once there is a slowdown of the number of constructions (like the one seen in UAE nowadays as compared to the previous decades), the number of laborers will reduce eventually because the infrastructure, facilities, and buildings are completed. This is why a good economic forecast is always tied with a forecast of population, land development, and planned growth of the infrastructure of a nation. Most master plans foresee 20 or 30 years into the future; some dare to plan for 50 years.

# Appendix 3

## *Case Study 2: Costa Rica Highways*

The effectiveness of maintenance interventions can be also tracked from other KPIs perspectives where decision-making is expanded to capture other key concerns.

## Effectiveness on Gas Emissions and Travel Time

In this case study, we estimate the effectiveness of travel time and GHG emissions by adding one additional lane to improve average speed on route 1 (Pan-American Highway) in Costa Rica.

The current speed ($V_{cur}$) is given by the free flow speed ($V_0$) reduced by a factor estimated by the level of traffic volume on the link ($Q$), the maximum capacity of the link ($Q_{max}$) of the equation calibrated in the exercises of Chap. 3.

$$V_{cur} = \frac{V_0}{1 + 0.3975 * \left(\frac{Q}{Q_{max}}\right)^{0.8033}} \tag{A.22}$$

We now proceed to estimate the before and after travel time on each segment for all multilane highways. First, notice the travel time before the expansion is simply the length of the segment divided by the current speed – $V_{cur}$ (adjusted to be meters per second). For segment 19090, it would be 881.0641 m divided by 12.388 m per second (44.60 km per hour divided by 3.6).

After the expansion, the volume per lane will drop; simply multiply the $Q$ before per lane by the number of lanes and divided over the number of lanes +1 additional lane. So for segment 19090, it will be 5499 = 6285 * 7 / (7 + 1).

Estimate the current speed using the calibrated equation shown before and using the $Q$ extra lane. For example, segment 19090 the $V_{cur}$ extra lane = 100 / (1 + 0.3975 * (5499 / 1524) ^ 0.8033) = 47.3

The after expansion travel time is then estimated by dividing the length of the segment over the $V_{cur}$ extra lane (in meters per second). For example, for segment 19090, it is 881.0641 meters / (47.3/3.6) = 67 seconds.

Consider now you are traveling on Route 1 (which has 10 segments); estimate the total travel time before and after: the summation of the travel time before is 5148.2 seconds and the after is 4644.3 seconds, or correspondingly 5148.2/60 = 85.8 minutes before and 77.4 minutes after for a 51.236 km segment (Table A.8).

Looking at route 1 only, estimate also the gas emissions; assume on average each vehicle produces 12 kg of $CO_2$ equivalent per hour (this is an illustrative figure), for route 1, segment 19002, the number of vehicles is 68,968; the before travel time is 259.67 seconds (0.072132 hours) and the after travel time is 235.289 seconds (0.065358 hours); the before and after kg of $CO_2$ are 4974.8 and 4507.6, a net saving of 467.2 kg (based on the assumed amount of 12 kg per hour per vehicle).

Similar calculations are applied for all segments, and a total reduction of 9.08 percent is observed between the 61,086 and the 55,542 kg for route 1 (Tables A.9 and A.10).

**Table A.8** Sample data for exercise

| Segment | Route | Lanes | Length | $V_0$ | $V_{cur}$ | $Q_{max}$ | $Q$ before per lane | $Q$ extra lane | $V_{cur}$ extra lane | Travel time before | Travel time after |
|---|---|---|---|---|---|---|---|---|---|---|---|
| 19090 | 39 | 7 | 881.1 | 100 | 44.60 | 1524 | 6285 | 5499 | 47.3 | 71 | 67 |
| 19003 | 1 | 6 | 3427.2 | 100 | 31.12 | 1717 | 14,568 | 12,487 | 34 | 396 | 365 |
| 19098 | 39 | 6 | 1542.7 | 100 | 28.81 | 1408 | 13,708 | 11,750 | 31 | 193 | 177 |
| 19103 | 39 | 6 | 1132.7 | 100 | 27.70 | 1239 | 12,927 | 11,080 | 30 | 147 | 135 |
| 20000 | 1 | 6 | 3072.4 | 100 | 33.44 | 1722 | 12,797 | 10,969 | 36 | 331 | 305 |
| 40710 | 1 | 6 | 3061.3 | 100 | 36.01 | 1754 | 11,312 | 9696 | 39 | 306 | 283 |
| 40040 | 1 | 6 | 3407.4 | 100 | 36.37 | 1780 | 11,258 | 9650 | 39 | 337 | 312 |
| 19100 | 39 | 6 | 1151.7 | 100 | 31.92 | 1361 | 11,032 | 9456 | 35 | 130 | 120 |
| 19099 | 39 | 6 | 725.0 | 100 | 30.06 | 1093 | 9868 | 8459 | 33 | 87 | 80 |
| 19097 | 39 | 6 | 494.8 | 100 | 26.16 | 784 | 9023 | 7734 | 29 | 68 | 62 |
| 19095 | 39 | 6 | 1179.4 | 100 | 40.19 | 1548 | 7998 | 6855 | 43 | 106 | 98 |
| 19101 | 39 | 6 | 1251.8 | 100 | 40.79 | 1579 | 7909 | 6779 | 44 | 110 | 103 |
| 19096 | 39 | 6 | 795.8 | 100 | 40.09 | 1394 | 7238 | 6204 | 43 | 71 | 66 |
| 19102 | 39 | 6 | 756.6 | 100 | 43.00 | 1435 | 6416 | 5499 | 46 | 63 | 59 |
| 19089 | 39 | 6 | 724.7 | 100 | 42.79 | 1413 | 6385 | 5473 | 46 | 61 | 57 |
| 19105 | 39 | 6 | 457.5 | 100 | 37.52 | 1072 | 6371 | 5461 | 40 | 44 | 41 |
| 19104 | 39 | 6 | 462.9 | 100 | 40.87 | 1164 | 5806 | 4976 | 44 | 41 | 38 |
| 19010 | 27 | 5 | 3207.7 | 100 | 28.54 | 1659 | 16,421 | 13,684 | 32 | 405 | 366 |
| 10070 | 27 | 5 | 3881.4 | 100 | 30.74 | 1741 | 15,100 | 12,584 | 34 | 454 | 412 |
| 19002 | 1 | 5 | 2213.5 | 100 | 30.69 | 1585 | 13,794 | 11,495 | 34 | 260 | 235 |
| 19024 | 104 | 5 | 2787.1 | 100 | 40.56 | 1784 | 9043 | 7536 | 44 | 247 | 227 |
| 30101 | 2 | 5 | 2029.6 | 100 | 40.83 | 1718 | 8587 | 7156 | 44 | 179 | 164 |
| 19011 | 32 | 5 | 1265.3 | 100 | 44.91 | 1648 | 6687 | 5573 | 49 | 101 | 94 |
| 19001 | 2 | 5 | 2189.7 | 100 | 49.85 | 1824 | 5777 | 4814 | 54 | 158 | 147 |
| 10003 | 2 | 5 | 7891.9 | 100 | 59.33 | 1983 | 1961 | 1634 | 63 | 479 | 452 |

(continued)

**Table A.8** (continued)

| Segment | Route | Lanes | Length | $V_0$ | $V_{cur}$ | $Q_{max}$ | $Q$ before per lane | $Q$ extra lane | $V_{cur}$ extra lane | Travel time before | Travel time after |
|---|---|---|---|---|---|---|---|---|---|---|---|
| 19094 | 39 | 4 | 973.0 | 90 | 10.82 | 557 | 21,063 | 16,850 | 13 | 324 | 278 |
| 10080 | 27 | 4 | 6240.0 | 90 | 27.39 | 1827 | 16,150 | 12,920 | 31 | 820 | 727 |
| 19093 | 39 | 4 | 616.9 | 90 | 12.07 | 445 | 14,391 | 11,513 | 14 | 184 | 159 |
| 19038 | 108 | 4 | 3039.7 | 90 | 29.76 | 1715 | 13,016 | 10,413 | 33 | 368 | 327 |
| 19026 | 105 | 4 | 1425.9 | 90 | 26.64 | 1395 | 12,948 | 10,358 | 30 | 193 | 171 |
| 19068 | 109 | 4 | 1343.4 | 90 | 28.61 | 1423 | 11,618 | 9295 | 32 | 169 | 150 |
| 19006 | 2 | 4 | 2257.4 | 90 | 31.21 | 1660 | 11,515 | 9212 | 35 | 260 | 232 |
| 19063 | 218 | 4 | 1180.3 | 90 | 28.34 | 1361 | 11,308 | 9047 | 32 | 150 | 133 |
| 19023 | 104 | 4 | 1498.4 | 90 | 30.92 | 1522 | 10,748 | 8598 | 35 | 174 | 156 |
| 30600 | 2 | 4 | 5115.6 | 90 | 34.33 | 1861 | 10,704 | 8563 | 38 | 536 | 482 |
| 19004 | 2 | 4 | 1979.1 | 90 | 32.51 | 1645 | 10,544 | 8435 | 36 | 219 | 196 |
| 19062 | 215 | 4 | 2290.2 | 90 | 33.11 | 1695 | 10,479 | 8383 | 37 | 249 | 223 |
| 20010 | 1 | 4 | 12087.3 | 90 | 35.90 | 1944 | 10,200 | 8160 | 40 | 1212 | 1092 |
| 19064 | 218 | 4 | 2295.0 | 90 | 33.78 | 1706 | 10,138 | 8110 | 38 | 245 | 219 |
| 19056 | 211 | 4 | 2829.2 | 90 | 34.49 | 1763 | 10,047 | 8037 | 38 | 295 | 265 |
| 19067 | 252 | 4 | 1636.7 | 90 | 33.15 | 1598 | 9861 | 7888 | 37 | 178 | 159 |
| 19005 | 2 | 4 | 2058.0 | 90 | 34.21 | 1684 | 9754 | 7803 | 38 | 217 | 194 |
| 19092 | 39 | 4 | 588.0 | 90 | 26.08 | 956 | 9212 | 7369 | 29 | 81 | 72 |
| 19017 | 100 | 4 | 1656.1 | 90 | 34.99 | 1636 | 9053 | 7243 | 39 | 170 | 153 |
| 19053 | 209 | 4 | 961.0 | 90 | 32.25 | 1378 | 8971 | 7177 | 36 | 107 | 96 |
| 19012 | 32 | 4 | 1845.8 | 90 | 36.38 | 1689 | 8620 | 6896 | 40 | 183 | 165 |
| 19020 | 102 | 4 | 773.3 | 90 | 32.43 | 1286 | 8283 | 6627 | 36 | 86 | 77 |
| 19037 | 108 | 4 | 358.7 | 90 | 18.96 | 494 | 8100 | 6480 | 22 | 68 | 59 |
| 30061 | 10 | 4 | 1518.7 | 90 | 37.18 | 1647 | 8032 | 6425 | 41 | 147 | 133 |
| 19052 | 209 | 4 | 1349.9 | 90 | 37.14 | 1611 | 7875 | 6300 | 41 | 131 | 118 |

| | | | | | | | | | | | |
|---|---|---|---|---|---|---|---|---|---|---|---|
| 19059 | 214 | 4 | 3108.6 | 90 | 40.09 | 1837 | 7596 | 6077 | 44 | 279 | 254 |
| 20250 | 153 | 4 | 1692.7 | 90 | 39.33 | 1709 | 7376 | 5901 | 43 | 155 | 141 |
| 19030 | 108 | 4 | 1604.0 | 90 | 39.61 | 1700 | 7219 | 5775 | 44 | 146 | 132 |
| 19047 | 204 | 4 | 1096.2 | 90 | 39.68 | 1591 | 6730 | 5384 | 44 | 99 | 90 |
| 19070 | 209 | 4 | 1313.0 | 90 | 40.60 | 1661 | 6673 | 5338 | 45 | 116 | 106 |
| 19065 | 218 | 4 | 3613.4 | 90 | 42.82 | 1877 | 6661 | 5329 | 47 | 304 | 277 |
| 19061 | 215 | 4 | 1661.0 | 90 | 41.62 | 1736 | 6585 | 5268 | 46 | 144 | 131 |
| 19050 | 207 | 4 | 1052.2 | 90 | 40.62 | 1595 | 6400 | 5120 | 45 | 93 | 85 |
| 19032 | 109 | 4 | 1104.2 | 90 | 41.32 | 1622 | 6260 | 5008 | 45 | 96 | 88 |
| 19091 | 39 | 4 | 870.2 | 90 | 41.11 | 1540 | 6010 | 4808 | 45 | 76 | 69 |
| 30040 | 10 | 4 | 2978.7 | 90 | 46.57 | 1879 | 5413 | 4330 | 51 | 230 | 212 |
| 19128 | 108 | 4 | 473.1 | 90 | 40.00 | 1261 | 5242 | 4193 | 44 | 43 | 39 |
| 19054 | 210 | 4 | 2109.8 | 90 | 47.23 | 1839 | 5106 | 4085 | 51 | 161 | 148 |
| 19039 | 167 | 4 | 2916.9 | 90 | 48.26 | 1887 | 4949 | 3959 | 52 | 218 | 201 |
| 19058 | 214 | 4 | 1512.5 | 90 | 47.59 | 1786 | 4861 | 3888 | 52 | 114 | 105 |
| 40750 | 3 | 4 | 2783.0 | 90 | 48.63 | 1884 | 4838 | 3870 | 53 | 206 | 190 |
| 40010 | 3 | 4 | 3616.9 | 90 | 49.24 | 1912 | 4747 | 3798 | 53 | 264 | 244 |
| 60621 | 17 | 4 | 3754.8 | 90 | 54.16 | 1936 | 3632 | 2906 | 58 | 250 | 233 |
| 10390 | 102 | 4 | 2916.5 | 90 | 54.53 | 1919 | 3525 | 2820 | 58 | 193 | 180 |
| 19025 | 104 | 4 | 4149.8 | 90 | 56.98 | 1950 | 3100 | 2480 | 61 | 262 | 246 |
| 19112 | 175 | 4 | 1436.9 | 90 | 58.08 | 1871 | 2782 | 2226 | 62 | 89 | 84 |
| 40610 | 112 | 4 | 421.2 | 90 | 61.67 | 1683 | 2001 | 1601 | 65 | 25 | 23 |
| 70410 | 240 | 4 | 2770.5 | 90 | 66.83 | 1966 | 1396 | 1117 | 70 | 149 | 143 |
| 51081 | 255 | 4 | 972.9 | 90 | 85.50 | 1997 | 48 | 38 | 86 | 41 | 41 |
| 60271 | 614 | 4 | 15109.7 | 90 | 87.09 | 2000 | 27 | 22 | 88 | 625 | 621 |
| 19016 | 100 | 3 | 1136.8 | 80 | 24.45 | 1321 | 11,584 | 8688 | 29 | 167 | 144 |
| 40000 | 3 | 3 | 4820.5 | 80 | 29.88 | 1847 | 11,083 | 8312 | 34 | 581 | 506 |

(continued)

**Table A.8** (continued)

| Segment | Route | Lanes | Length | $V_0$ | $V_{cur}$ | $Q_{max}$ | $Q$ before per lane | $Q$ extra lane | $V_{cur}$ extra lane | Travel time before | Travel time after |
|---|---|---|---|---|---|---|---|---|---|---|---|
| 10450 | 147 | 3 | 2487.5 | 80 | 29.58 | 1718 | 10,520 | 7890 | 34 | 303 | 263 |
| 19019 | 102 | 3 | 1747.6 | 80 | 29.06 | 1611 | 10,210 | 7658 | 33 | 216 | 188 |
| 19015 | 100 | 3 | 354.8 | 80 | 7.45 | 179 | 9690 | 7268 | 9 | 171 | 140 |
| 10211 | 209 | 3 | 1347.9 | 80 | 32.25 | 1595 | 8188 | 6141 | 37 | 150 | 132 |
| 20020 | 1 | 3 | 7061.7 | 80 | 36.40 | 1928 | 7596 | 5697 | 41 | 698 | 619 |
| 20031 | 1 | 3 | 7980.8 | 80 | 36.72 | 1938 | 7478 | 5609 | 41 | 782 | 694 |
| 19034 | 110 | 3 | 2507.8 | 80 | 37.41 | 1821 | 6731 | 5048 | 42 | 241 | 215 |
| 19076 | 200 | 3 | 548.3 | 80 | 30.87 | 1188 | 6677 | 5008 | 35 | 64 | 56 |
| 20132 | 3 | 3 | 2242.0 | 80 | 37.64 | 1804 | 6576 | 4932 | 42 | 214 | 191 |
| 20131 | 3 | 3 | 2537.4 | 80 | 38.17 | 1831 | 6450 | 4838 | 43 | 239 | 213 |
| 20050 | 1 | 3 | 4772.7 | 80 | 39.35 | 1913 | 6262 | 4696 | 44 | 437 | 390 |
| 20040 | 1 | 3 | 4241.8 | 80 | 39.30 | 1902 | 6247 | 4686 | 44 | 389 | 347 |
| 19121 | 177 | 3 | 788.0 | 80 | 37.13 | 1517 | 5707 | 4280 | 42 | 76 | 68 |
| 21430 | 27 | 3 | 14001.2 | 80 | 35.98 | 1973 | 5664 | 4248 | 41 | 1401 | 1241 |
| 21890 | 27 | 3 | 15214.7 | 80 | 36.48 | 1976 | 5486 | 4114 | 41 | 1501 | 1332 |
| 40180 | 113 | 3 | 1177.7 | 80 | 40.73 | 1709 | 5133 | 3850 | 45 | 104 | 93 |
| 10920 | 251 | 3 | 1791.8 | 80 | 41.76 | 1810 | 5099 | 3824 | 46 | 154 | 139 |
| 60623 | 17 | 3 | 3812.2 | 80 | 43.48 | 1915 | 4843 | 3632 | 48 | 316 | 286 |
| 60180 | 27 | 3 | 4899.9 | 80 | 47.06 | 1947 | 3919 | 2939 | 51 | 375 | 343 |
| 21442 | 27 | 3 | 6822.0 | 80 | 47.85 | 1963 | 3823 | 2868 | 52 | 513 | 470 |
| 19048 | 205 | 3 | 3259.1 | 80 | 49.65 | 1933 | 3285 | 2464 | 54 | 236 | 218 |
| 70160 | 32 | 3 | 10274.3 | 80 | 41.87 | 1982 | 2797 | 2098 | 46 | 883 | 796 |
| 10950 | 32 | 3 | 5011.7 | 80 | 52.56 | 1963 | 2790 | 2093 | 57 | 343 | 318 |
| 21441 | 27 | 3 | 11639.2 | 80 | 41.56 | 1990 | 1711 | 1284 | 46 | 1008 | 907 |
| 70730 | 257 | 3 | 1932.4 | 80 | 77.07 | 1999 | 32 | 24 | 78 | 90 | 90 |

**Table A.9** Gas emissions reduction

| Section | RUTA | Lanes | Shape_Length | Forecasted speed | $Q$ | $Q$ extra lane | Vcur extra lane | Travel time before | Travel time after | Hours before | Hours after | Cars | Kg $CO_2$ before | Kg $CO_2$ after |
|---|---|---|---|---|---|---|---|---|---|---|---|---|---|---|
| 19002 | 1 | 5 | 2213.457 | 30.6862 | 13793.6 | 11494.67 | 33.86663 | 259.6752 | 235.289 | 0.072132 | 0.065358 | 68,968 | 4974.8 | 4507.615 |
| 19003 | 1 | 6 | 3427.223 | 31.12089 | 14567.83 | 12486.71 | 33.81763 | 396.4541 | 364.8394 | 0.110126 | 0.101344 | 87,407 | 9625.795 | 8858.198 |
| 20000 | 1 | 6 | 3072.384 | 33.44498 | 12796.83 | 10968.71 | 36.2473 | 330.7098 | 305.1422 | 0.091864 | 0.084762 | 76,781 | 7053.396 | 6508.09 |
| 20010 | 1 | 4 | 12087.33 | 35.9041 | 10199.75 | 8159.8 | 39.85087 | 1211.962 | 1091.931 | 0.336656 | 0.303314 | 40,799 | 13735.23 | 12374.91 |
| 20020 | 1 | 3 | 7061.667 | 36.39904 | 7596.333 | 5697.25 | 41.04583 | 698.425 | 619.3564 | 0.194007 | 0.172043 | 22,789 | 4421.224 | 3920.698 |
| 20031 | 1 | 3 | 7980.759 | 36.72437 | 7478 | 5608.5 | 41.37463 | 782.3343 | 694.4046 | 0.217315 | 0.19289 | 22,434 | 4875.246 | 4327.298 |
| 20040 | 1 | 3 | 4241.78 | 39.29633 | 6247.333 | 4685.5 | 43.95212 | 388.5962 | 347.4328 | 0.107943 | 0.096509 | 18,742 | 2023.075 | 1808.774 |
| 20050 | 1 | 3 | 4772.747 | 39.34961 | 6261.667 | 4696.25 | 44.0051 | 436.647 | 390.4522 | 0.121291 | 0.108459 | 18,785 | 2278.448 | 2037.401 |
| 40040 | 1 | 6 | 3407.389 | 36.36798 | 11258.33 | 9650 | 39.28336 | 337.2913 | 312.2594 | 0.093692 | 0.086739 | 67,550 | 6328.896 | 5859.202 |
| 40710 | 1 | 6 | 3061.28 | 36.00811 | 11311.67 | 9695.714 | 38.91076 | 306.0591 | 283.2278 | 0.085016 | 0.078674 | 67,870 | 5770.064 | 5339.631 |
| | | | | | | | | | | | | Total | 61086.17 | 55541.82 |
| | | | | | | | | | | | | Reduction | | 5544.357 |

**Table A.10** Treatment effectiveness of safety

| Section | Length | V85 before | V85 after | Ln (*Q*) before | Ln (*Q*) after | Crash before per km | Crash after per km | Total crashes | Total crashes |
|---|---|---|---|---|---|---|---|---|---|
| 19002 | 2213.5 | 30.7 | 33.9 | 9.2 | 8.8 | 35.0 | 34.0 | 77.4 | 75.4 |
| 19003 | 3427.2 | 31.1 | 33.8 | 8.8 | 8.5 | 33.4 | 32.6 | 114.3 | 111.7 |
| 20000 | 3072.4 | 33.4 | 36.3 | 8.8 | 8.5 | 33.9 | 33.2 | 104.3 | 102.0 |
| 20010 | 12087.3 | 35.9 | 39.9 | 7.2 | 6.7 | 27.3 | 26.1 | 330.1 | 315.4 |
| 20020 | 7061.7 | 36.4 | 41.1 | 7.5 | 6.8 | 28.6 | 26.7 | 201.7 | 188.7 |
| 20031 | 7980.8 | 36.7 | 41.4 | 7.3 | 6.6 | 28.0 | 26.2 | 223.6 | 208.9 |
| 20040 | 4241.8 | 39.3 | 44.0 | 7.8 | 7.1 | 30.8 | 29.0 | 130.8 | 123.0 |
| 20050 | 4772.8 | 39.4 | 44.0 | 7.7 | 7.0 | 30.3 | 28.5 | 144.7 | 135.9 |
| 40040 | 3407.4 | 36.4 | 39.3 | 8.6 | 8.2 | 33.7 | 33.0 | 114.9 | 112.5 |
| 40710 | 3061.3 | 36.0 | 38.9 | 8.7 | 8.4 | 34.1 | 33.4 | 104.5 | 102.3 |

| Travel time before | Travel time after | Hour before | Hours after | Cars | Kg $CO_2$ before | Kg $CO_2$ after |
|---|---|---|---|---|---|---|
| 259.675 | 235.289 | 0.072 | 0.065 | 68,968 | 4974.8 | 4507.615 |

### Capturing Treatment Effectiveness of Safety

For route 1, assume we have a safety performance model given by the following equation:

$$\text{Frequency} = 4.59319^* \ \ln(\text{PC\_MI\_LN\_HR}) + 0.28776^*\text{Speed\_V85} - 16.16$$

Assume now you add TWO more lanes and wish to know how much road collisions would reduce on the segments of route 1.

***Solution*** We take the before and after volume (*Q*) in passenger car per mile, per lane, per hour and use it to estimate the improved speed and reductions in crashes per km and then in total with the previous equation.

## References


Huang, Y. (2003). *Pavement analysis and design* (2nd ed.).

ITE. (2017). *Trip generation handbook* (3rd ed.). Institute of Transportation Engineers.

De la Barra. (2014). TRANUS User Guide. TRANUS : Integrated Land Use and Transport Modelling System.

NCHRP. (2004). Guide for mechanistic-empirical design of new and rehabilitated pavement structures. Appendix GG-1: Calibration of permanent deformation models for flexible pavements.

Societe de Transport de Montreal. (2012). Strategic Plan. 2020.

TDB. (2018). Trips Database Bureau User Guide.

TRICS. (2018). TRICS multi-modal methodology. TRICS Consortium Limited, by Ian Coles.

# Index





A. Mohammadi, L. Amador Jimenez, *Asset Management Decision-Making For Infrastructure Systems*, https://doi.org/10.1007/978-3-030-97614-9

The manufacturer's authorised representative in the EU is Springer Nature Customer Service Centre GmbH, Europaplatz 3, 69115 Heidelberg, Germany. If you have any concerns regarding our products, please contact ProductSafety@springernature.com

Printed and bound by CPI Group (UK) Ltd, Croydon, CR0 4YY
30/09/2026
02225265-0001